Radio Amateurs' Examination Manual

FOURTEENTH EDITION

George Benbow, G3HB

with

Ian Jackson, G3OHX

Clive Smith, G4FZH

Radio Society of Great Britain

Published by the Radio Society of Great Britain, Cranborne Road, Potters Bar, Herts EN6 3JE.

Fourteenth edition 1993
Reprinted with corrections 1994

ISBN 1 872309 25 9

Cover design: Geoff Korten Design.
Cover photos: Thomas Neile Photographers in association with ICOM (UK) Ltd.
Illustrations: Derek Cole, Radio Society of Great Britain.
Production and typography: Ray Eckersley, Seven Stars Publishing.

Printed in Great Britain by Bell & Bain Ltd, Glasgow.

Contents

Preface

Regulations for the UK amateur licence are now deleted from this manual and the RAE candidate should therefore obtain the current edition (Revision 3, March 1994) of *How to Become a Radio Amateur* and also *Amateur Radio Licence (A or B) Terms, Provisions and Limitations Booklet BR68* for the appropriate information. A number of other amendments, deletions and changes in presentation have also been made. In order that CEPT countries may issue licences to UK amateurs, reference to valves as power amplifiers is now included.

Thanks are due to the City and Guilds of London Institute for permission to reproduce the RAE syllabus and objectives in Appendix 3, and certain specimen RAE questions which have been included in Appendix 4.

The permission of the Department of Trade and Industry to reproduce information from *How to Become a Radio Amateur* and also from *Free Space Propagation* Technical Instruction 05/86 (V R Brown, Radio Investigation Service) is also gratefully acknowledged.

Figs 7.3–7.6 inclusive are taken from*Short Wave Radio and the Ionosphere* by Bennington, 2nd edition, and are reproduced here by permission of *Wireless World*.

The *RAE Manual* is intended primarily as a basis for formal tuition and contains all the technical information necessary to pass the examination, although the treatment is necessarily brief. It is difficult to recommend a more elementary introduction to basic electronics, if needed, to meet the requirements of individual private students. A visit to the local library is therefore suggested!

Other books published by the RSGB are of value to RAE students. *How to Pass the RAE* explains the background to the multiple-choice type of examination. It also contains more sample examination papers (with answers) in the RAE format. *RAE Revision Notes* is a summary of the salient points of the *Manual* and is intended for revision.

George Benbow, G3HB

CHAPTER 1

Becoming a radio amateur

Before an amateur radio station can be established and used, it is necessary to obtain a licence from the Secretary of State for Trade and Industry.

Full details are given in the publication *How to Become a Radio Amateur* which is obtainable, free of charge, from Amateur Radio Licensing Section, Radiocommunications Agency, Room 712, Waterloo Bridge House, Waterloo Road, London SE1 8UA.

The necessary requirements are summarised in this chapter.

Requirements of the Amateur Licence

Amateur Licence A

1. Applicants must be over 14 years of age.
2. Applicants must have passed the Radio Amateurs' Examination (RAE).
3. Applicants must have passed the Morse Test.
4. The appropriate fee must be paid before the licence is issued and each year on the anniversary of the date of issue.

Amateur Licence B

This licence does not authorise the use of frequencies below 30MHz. The use of morse telegraphy is permitted but identification of the station must be by telephony, otherwise its conditions are broadly the same as those of the Amateur Licence A except that one will not be required to pass the morse test. Because of the requirement for station identification by telephony, operation must be in the all-mode segment of the bands and not in the telegraphy-only segment.

Commonwealth or alien citizens who reside in the UK may take the RAE and obtain the UK licence.

The Radio Amateurs' Examination

The City and Guilds of London Institute, 76 Portland Place, London W1N 4AA holds the Radio Amateurs' Examination, usually in May and December. This can be taken at local colleges and examination centres throughout the country. Application to sit the examination should be made well in advance so that the college is aware in good time of the need to arrange for this examination to be held there. Closing dates for applications to sit the examinations may be obtained from the colleges.

The regulations, syllabus and objectives for the RAE can be obtained from the City and Guilds of London Institute at the above address for a small charge.

The syllabus and objectives are given in Appendix 3 of this book, and practice multiple-choice questions in Appendix 4.

The examination objectives indicate in broad terms what the candidate is expected to know and the syllabus defines the scope of the questions he or she will be asked in the examination.

The examination lasts for three hours.

Paper 1 on licensing conditions, transmitter interference and electromagnetic compatibility takes one hour and 15 minutes and consists of 45 multiple-choice questions. Each subject has 15 questions.

There is then a break of 15 minutes.

Paper 2 takes one hour 30 minutes and has 55 multiple-choice questions. These questions are apportioned as follows:

Subject	Questions
1. Operating procedures	9
2. Electrical theory	6
3. Solid-state devices	7
4. Receivers	7
5. Transmitters	8
6. Propagation and antennas	9
7. Measurements	9
Total	55

Candidates must take both components (papers) on their first entry. Candidates who are successful in one but not both components may carry forward their success and need subsequently retake only the component in which they were unsuccessful.

Details of colleges and institutions which offer courses leading to the RAE are given in *Radio Communication*, the journal of the Radio Society of Great Britain, and other radio magazines, usually in July, August and September each year.

Tuition and examination fees vary from college to college, and such information may be obtained from the college on enrolment.

Candidates must obtain from the Radiocommunications Agency the publications referred to in the first paragraph of the Preface.

Candidates are strongly recommended to read the guide *How to Improve Television and Radio Reception* which is published by the Department of Trade and Industry and is obtainable free of charge from the Radiocommunications Agency Library, telephone 071-215 2352.

A pass in the Radio Amateurs' Examination is regarded as valid for life.

The Morse Test

A list of centres available, with dates and times of tests and an application form may be obtained from Morse Tests, Amateur Radio Dept, Radio Society of Great Britain, Lambda House, Cranborne Road, Potters Bar, Herts EN6 3JE.

The Amateur Radio Morse Test now consists of a receiving test, sent on a manual key, and a sending test, also sent on a manual key. Both tests are based on typical exchanges between radio amateurs. The receiving test lasts for approximately 2½ minutes with a maximum of six uncorrected errors; the sending test lasts for approximately 1½ minutes with not more than four corrected and no uncorrected errors. Full details are given in *The Morse Code for Radio Amateurs* published by the RSGB.

A pass in the Amateur Radio Morse Test is regarded as valid for life.

It should be noted that the requirement to pass a test in morse code is laid down by the International Telecommunications Union.

Either the Morse Test or the RAE may be taken first.

The Amateur Radio Certificate

The Amateur Radio Certificate is no longer issued but this does not affect the validity of those certificates which have been issued.

The holder of an ARC should quote the number of the certificate when applying for the Amateur Licence (A).

A previously held licence may be re-issued to the legitimate holder (even when the original qualifications were not based upon the current C & G RAE syllabus). Details of the procedure for the issue of lapsed licences can be obtained from Amateur Radio Licensing Section, Radiocommunications Agency, Room 712, Waterloo Bridge House, Waterloo Road, London SE1 8UA.

Fees

The fee, on issue and on annual renewal, for Amateur Licence A and Amateur Licence B is currently (1994) £15.

Background knowledge

It is most important that the RAE candidate should acquire as much background knowledge of amateur radio as possible. A period as a shortwave listener is particularly valuable; one then becomes familiar with amateur radio communication, how propagation governs which part of the world can be heard and when and on which waveband, operating procedures and so on. There are a number of magazines devoted to amateur radio available, including *Radio Communication* published by the RSGB. While every article in these is obviously not aimed at the beginner, much useful information can be found in them. If there is a local radio society, join it! There, one meets other amateurs, some with wide experience and some just raw beginners. Talking to them and listening to their conversations can be most useful.

Too often one hears expressed the erroneous belief that the conditions imposed by the authorities have been devised to discourage the experimenter. Such is very far from the truth: the newcomer can be confident that his or her desire to obtain a licence will meet with courtesy, assistance and every encouragement from the authorities – provided that no special concessions are expected.

Reciprocal licences

There are now agreements between the UK and certain foreign countries whereby citizens of those countries who hold a transmitting licence issued by their own government may obtain a UK licence and vice versa.

Details of such licences are outside the scope of this book and intending applicants for reciprocal licences should seek guidance from the Radio Society of Great Britain.

Temporary permission

The UK licence now permits holders to operate as temporary visitors in countries which have implemented the CEPT recommendation T/R 61-01.

The Radio Society of Great Britain

Every reader of this manual is advised to become, if not already so, a member of the Radio Society of Great Britain (RSGB).

The RSGB is the national society for radio amateurs in the UK. The majority of its members hold amateur transmitting licences: the others either hope to do so later or are interested primarily in the receiving side of amateur radio.

The Society acts as the spokesman for the radio amateur and amateur radio in the UK, and is one of the founder members of the International Amateur Radio Union, the worldwide association of the various national societies.

The Society was founded as the London Wireless Club in 1913 but soon attracted members throughout the country. The name 'Radio Society of Great Britain' was formally adopted in 1922. For many years its activities have been devoted almost entirely to the many aspects of amateur radio, that is, the transmission and reception of radio signals as a hobby pursued for the pleasure to be derived from an interest in radio techniques and construction and for the ensuing friendships with like-minded persons throughout the world.

The Society is recognised as the representative of the amateur radio movement in all negotiations with the

Department of Trade and Industry (DTI) on matters affecting the issue of amateur transmitting licences.

The Society maintains close liaison with the DTI on all matters affecting licence facilities and the frequencies assigned to amateur radio, and sends official representatives to the important World Administrative Radio Conferences of the International Telecommunication Union and other conferences where decisions vital to the future of amateur radio are taken.

The Society helps amateur radio in many ways. Of particular importance is the provision of information on technical matters and on the various activities and events of concern to amateurs. Since 1925 it has published a monthly journal, *Radio Communication*, the oldest and largest magazine in this country devoted to amateur radio. All members receive this magazine by post, without payment other than their annual membership subscriptions.

Anyone over 18 years of age or holding an amateur transmitting licence is eligible to become a Corporate Member of the Society. It is not necessary to be engaged professionally in radio but equally this would not debar anyone from joining. Many members do in fact work in the electronics field, but for very many others radio is purely a spare-time hobby. If you are over 65 years of age or a student under the age of 25 who holds a UK licence, you are eligible to become a Corporate (concessionary) member at a reduced subscription. Those under 18 years old who do not hold an amateur transmitting licence may become Associate members. Associates have many of the privileges of full membership but do not vote in the annual Council election or on matters affecting the management of the Society. Associates must apply for transfer to Corporate membership on reaching 18 years of age or immediately they obtain a transmitting licence if under this age.

Full details of the aims, activities and advantages of membership of the Radio Society of Great Britain may be obtained from its Headquarters at Lambda House, Cranborne Road, Potters Bar, Herts EN6 3JE.

CITY AND GUILDS RADIO AMATEURS' EXAMINATION (765)
EXAMINING CANDIDATES WITH SPECIAL NEEDS

City and Guilds of London Institute has for many years made arrangements for candidates with a wide range of special needs, who would have difficulties taking the Radio Amateurs' Examination under normal conditions. Essentially, those who are housebound, blind, or handicapped in some other way, may be examined at home. Moreover, if the candidate is unable to complete a written paper, it is possible for an examination to be conducted orally. (This provision makes radio amateurs unique – it is the only one of City and Guilds' schemes which can be examined in this way.)

Candidates who wish to take advantage of these special arrangements will be required to provide some official notification which confirms the reason they have given for requesting home examination. Examples of acceptable documentation include a doctor's certificate or a DHSS letter. In all cases, however, the final decision concerning acceptance (or otherwise) rests with City and Guilds.

Application forms are available from Section 12, City and Guilds, 45 Britannia Street, London WC1X 9RG; further information is also available from the same source. Please note that completed application forms must be received at City and Guilds by 25 October (for examination in December) or 14 March (for the May examination).

Candidates must also be prepared to make themselves available for examination in the 10 days following the date of the regular examination. (City and Guilds of London Institute.)

CHAPTER 2

Basic radio theory, circuits and calculations

Molecules, atoms, and electrons

All matter is comprised of molecules; the molecule is the smallest quantity of a substance which can exist and still display all the physical and chemical properties of that substance.

Molecules are made up of smaller particles called 'atoms' of which there are over 100 different types. All molecules consist of various combinations of these atoms, for example, two atoms of hydrogen and one of oxygen form one molecule of water, and sulphuric acid is made up of two atoms of hydrogen, one of sulphur and four of oxygen.

Atoms are so small that they cannot be seen under the most powerful microscope but their constitution is of vital importance in electrical and communication engineering.

Atoms consist of a relatively heavy, positively charged core or nucleus, around which a number of much lighter negatively charged electrons move in one or more orbits.

One type of atom differs from another in the number of positive and neutral particles known as 'protons' and 'neutrons' which make up the nucleus and the number and arrangement of the negative 'electrons' which are continually orbiting round the nucleus. Some atoms are extremely complex, having a large number of electrons in several orbits, and others are quite simple.

Under some circumstances it is possible to detach an electron from an atom, particularly when its outer orbit contains only one electron. In other atoms it is virtually impossible to detach an electron.

Conductors and insulators

The ease with which electrons can be detached from their parent atoms thus varies from substance to substance. In some substances there is a continual movement of electrons in a random manner from one atom to another, and the application of a voltage (for example from a battery) to the two ends of a piece of wire made of such a substance will cause a drift of electrons along the wire – this is an 'electric current'. It should be noted that if an electron enters the wire from the battery at one end it will be a different electron which immediately leaves the other end of the wire.

By convention, the direction of current flow is said to be from positive to negative, hence the term 'conventional current'. Materials which conduct electricity are called 'conductors'. All metals belong to this class. Materials which do not conduct electricity are called 'insulators'. See Table 2.1.

The utilisation of electricity in all branches of electrical engineering depends on the existence of a conductor to carry the electric current and insulators to restrict the flow of the current to within the conductor.

In practice, the conductor is almost universally a single or stranded wire of copper, generally tinned for ease of soldering. For some applications the copper may be silver plated.

Many different insulators are in common use, from mica and glazed ceramic to synthetic materials such as PVC, polythene, polystyrene and PTFE, the last three of which have good insulating properties at very high frequencies.

Electrical units

Current

The unit of current flow is called the 'ampere' (amp) and the strength of a current is said to be '*x* amperes'. Currents are denoted in formulae by the symbol *I*. The currents used in radio are often very small fractions of an ampere and for convenience the two small units 'milliampere' (10^{-3}A) and 'microampere' (10^{-6}A) are used. Thus a current of 0.003 ampere is written as '3 milliamperes'. See Table 2.2 for abbreviations.

Voltage

In order to make a current flow through a circuit, it is necessary to have some device which can produce a continuous supply of electrons. This may be a battery, in which the supply of electrons is produced by chemical action, or a dynamo or generator in which mechanical energy is turned

Table 2.1. Materials commonly used as conductors and insulators

Conductors	Insulators
Silver	Mica
Copper	Ceramics
Aluminium	Plastics

Table 2.2. Units and symbols

Quantity	Symbol used in formulae	Unit	Abbreviation
current	I	ampere	A
EMF	E	volt	V
electric potential	V	volt	V
time	t	second	s
resistance	R	ohm	Ω
capacitance	C	farad	F
inductance	L	henry	H
mutual inductance	M	henry	H
power	W	watt	W
frequency	f	hertz (one cycle per second)	Hz
wavelength	λ	metre	m

Abbreviations for multiples and sub-multiples

G	**giga**	10^{9}
M	**mega**	10^{6}
k	**kilo**	10^{3}
c	**centi**	10^{-2}
m	**milli**	10^{-3}
µ	**micro**	10^{-6}
n	**nano**	10^{-9}
p	**pico**	10^{-12}

into electrical energy. The battery or generator produces an 'electromotive force' (EMF, symbol E) which may be used to force a current through a circuit. The unit of electric potential is the 'volt', and voltages are usually denoted in formulae by the symbol V.

Resistance

The ease with which an electric current flows through a wire depends on the dimensions of the wire and the material from which it is made. The opposition of a circuit to the flow of current is called the 'resistance' (R) of the circuit. The resistance of a circuit is measured in 'ohms' (Ω). For convenience, because the resistances used in radio equipment may be up to 10,000,000Ω, two larger units called the 'kilohm' (1000Ω) and the 'megohm' (1,000,000Ω) are used. Thus 47,000Ω may be abbreviated to 47kΩ.

The direct current circuit

Fig 2.1 is the simplest possible circuit, a current from a battery flows through R. The ratio of the voltage across the circuit to the current which flows though it is a constant:

$$R = \frac{V}{I}$$

which is known as the 'resistance' (R), and is the opposition to the flow of the electric current, while the relationship is known as 'Ohm's Law'; V is measured in volts and I in amperes; R is then in ohms.

It should be noted that in this circuit the current I also flows through the battery,which has an internal resistance r. Thus the EMF of the battery is the total voltage available to drive the current I through a total resistance of $(R + r)$. The EMF is thus equal to $(IR + Ir)$. Some voltage is inevitably lost in driving the current through the battery itself. That which is left to do useful work is known as the 'potential difference' between the points a and b. The best battery is therefore one with the lowest internal resistance.

Fig 2.1. Simple DC circuit

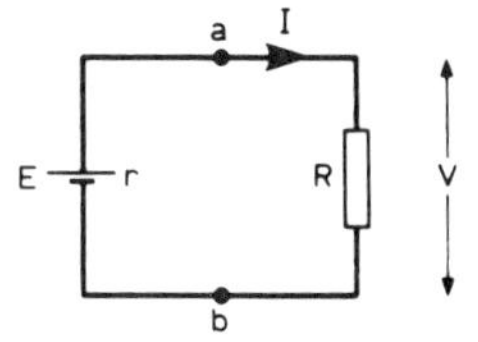

Power in the DC circuit

The passage of an electric current through a resistance causes heat to be dissipated in the resistance. Thus electrical energy is converted into heat.

The power dissipated in the resistance is:

$$\text{power (watts)} = \text{voltage (volts)} \times \text{current (amps)}$$
$$W = V \times I$$

By the use of Ohm's Law, the power dissipated in the resistance may be expressed in two other forms:

$$W = V^2/R \quad \text{and} \quad W = I^2 \times R$$

All materials have the property of resistance. In the case of metals suitable for use as conductors of electricity, eg copper (silver is better but is of course much more expensive), it is very low. Special alloys intended for heating elements are made with a very high resistance. Nichrome, for example, has a resistance which is about 60 times that of copper. Insulators are materials which have an extremely high resistance and therefore for all practical purposes do not conduct electricity. Some materials, eg germanium and silicon, have a resistance which is higher than that of conductors but is lower than insulators. These are known as 'semiconductors'.

The resistance of a conductor is proportional to its length and inversely proportional to its cross-sectional area. It also depends upon the material from which the conductor is made.

The 'resistivity' of a material is the resistance measured between the opposite faces of a 1cm cube of the material.

Resistors

In series

A discrete component having the property of resistance is called a 'resistor'. A number of these can be connected as shown in Fig 2.2(a). This is the 'series' connection and the effective resistance, R, is:

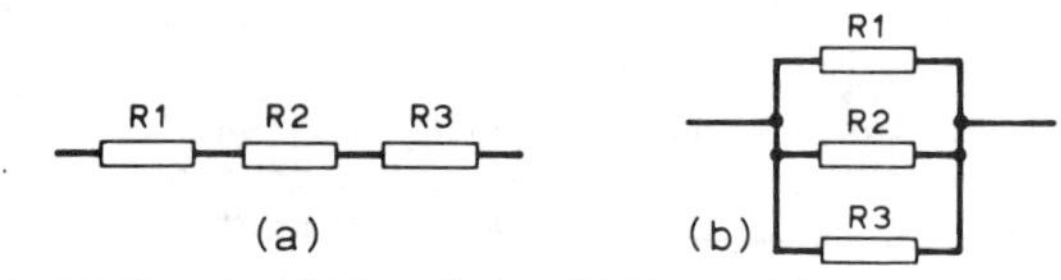

Fig 2.2. Resistors (a) in series, and (b) in parallel

$$R = R_1 + R_2 + R_3 + \ldots$$

In parallel

The 'parallel' connection is shown in Fig 2.2(b). The effective resistance in this case is:

$$\frac{1}{R} = \frac{1}{R_1} + \frac{1}{R_2} + \frac{1}{R_3} + \ldots$$

or

$$R = \frac{1}{\frac{1}{R_1} + \frac{1}{R_2} + \frac{1}{R_3} + \ldots}$$

The effective resistance of only two resistors in parallel is:

$$R = \frac{R_1 \times R_2}{R_1 + R_2} \quad \text{ie} \quad \frac{\text{Product}}{\text{Sum}}$$

The commonest forms of resistor are:

(a) carbon (in the form of a rod);
(b) spiral carbon or metal oxide track (on glass or ceramic former);
(c) wire wound (with high-resistance wire).

They are graded according to their 'dissipation', ie the amount of heat they can dissipate safely for a given temperature rise.

A connection may be taken from the junction of two resistors in series. This combination is known as a 'fixed potentiometer' or 'potential divider' because when it is connected across a source of voltage it enables any required proportion of the voltage to be obtained according to the values of the two resistors. These are generally fairly high in value to avoid putting too heavy a load on the source. The two resistors are often replaced by a single resistor in which the position of the 'tap' is varied by a sliding contact (a 'variable potentiometer').

The main application of the resistor in electronic circuits is to create a given voltage drop across it for a particular purpose as a result of a known current flowing through the resistor (by Ohm's Law). A fairly high value of resistor may often be employed to provide a leakage path to earth from a particular part of the circuit. Resistors are also used as the load across which the output of an amplifier stage is developed, or in which the output of a transmitter is dissipated.

The alternating current circuit

In the AC circuit the voltage and current are not constant with time as in the DC circuit; the value of each alternates between positive and negative states.

The AC waveform is shown in Fig 2.3 and is a 'sinewave' or 'sinusoidal' waveform. There are two values of the amplitude of this waveform which are relevant:

(a) the peak value;
(b) the RMS value.

The peak value is clear from Fig 2.3 and the root mean square (RMS) value is that value which is equivalent in heating effect to a DC supply of the same value. For a sine wave, the RMS value is 0.707 times the peak value. The RMS value is used to define an alternating voltage, ie the standard 50Hz supply mains is 240V (RMS) (the peak value is therefore 340V).

Two other values of use are the 'average' value, which is 0.636 times the peak value, and the 'instantaneous' value, which is the value of the current (or voltage) at a particular instant in an alternating cycle. It is usually denoted by small letters, ie i (or v).

The time occupied by one complete cycle is the 'period T' and the number of cycles per second is the 'frequency f'. Thus:

$$f = \frac{1}{T} \quad \text{and} \quad T = \frac{1}{f}$$

Phase difference between waveforms

Phase in this context means 'time' or time difference between two waveforms. For convenience, this time difference or phase difference is measured in degrees; one complete cycle of the waveform is taken to be 360° and a half cycle is 180° etc. Thus the time difference between two alternating waveforms can be defined by the phase angle between them.

Two alternating waveforms are said to be 'in phase' when they begin at the same point in time: see Fig 2.4(a). At any other point they are 'out of phase'. The term 'in phase opposition' is sometimes used to describe a phase difference of 180°. In this case, two waveforms of equal amplitude would cancel each other.

In Fig 2.4(b), A leads B by 90°, conversely B lags on A by 90°.

Distortion of alternating waveforms

Distortion of an alternating waveform is caused by the presence of other sinusoidal waveforms of frequencies which are related to the original frequency (known as the 'fundamental');

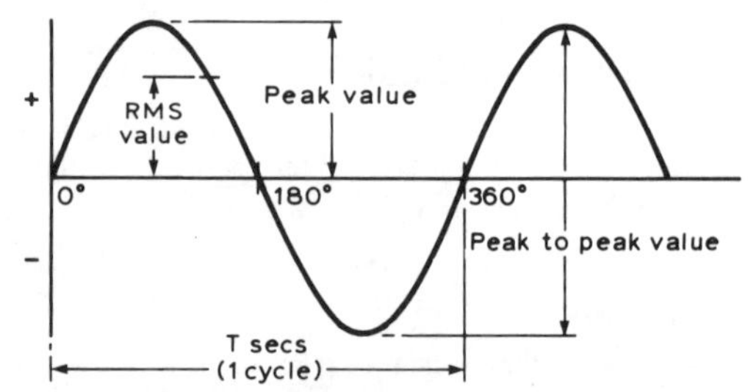

Fig 2.3. Alternating (sinusoidal) waveform

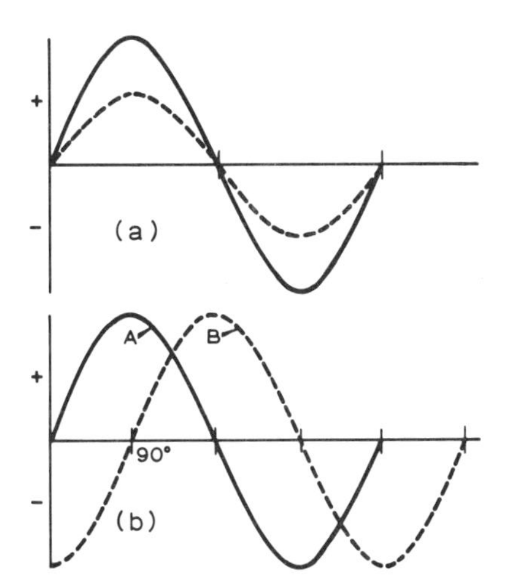

Fig 2.4. (a) Two alternating waveforms in phase, ie they start at the same point in time. (b) Two alternating waveforms with a phase difference of 90°

thus, if the fundamental frequency is f, then $2f$ is the 'second harmonic', $3f$ is the 'third harmonic' and so on.

Fig 2.5 shows the distortion resulting from the addition of 30% of 2nd harmonic to the fundamental (a) and 20% of 3rd harmonic to the fundamental (b). Distortion increases as the number and amplitude of the harmonics present increase. A distorted waveform is often known as a 'complex' waveform.

Inductance and capacitance in the AC circuit

Two new circuit elements are of great significance in the AC circuit. These are:

(a) the inductor which has inductance;
(b) the capacitor which has capacitance.

A circuit possesses inductance if it can store energy in the form of a magnetic field. The unit of inductance is the henry (H) and the symbol for inductance is L. A circuit has an inductance of one henry if a current in it, changing at the rate of one ampere per second, induces an EMF of one volt. The energy stored in an inductor is $\frac{1}{2}LI^2$ joules, where L is in henrys and I is in amperes.

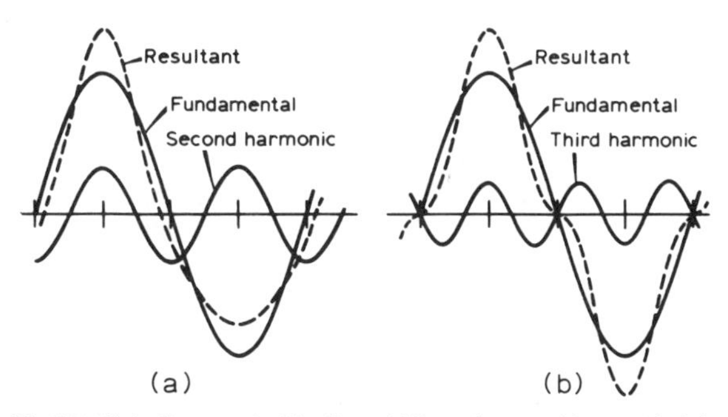

Fig 2.5. Distortion created by the addition of second harmonic (a) and third harmonic (b) to fundamental

A circuit possesses capacitance if it can store energy in the form of an electric field. The unit of capacitance is the farad (F); the symbol for capacitance is C. A circuit has a capacitance of one farad if a charge of one coulomb sets up a voltage of one volt across it. The energy stored in a capacitor is $\frac{1}{2}CV^2$ joules, where C is in farads and V is in volts. The farad is an impracticably large unit and the practical unit is the 'microfarad' or µF (0.000001 farad).

Note that no energy is stored in an inductor if there is no current flowing. In a capacitor, however, there need be no movement of charge and the energy stored is static. A good-quality capacitor can maintain a considerable, perhaps lethal, voltage across its terminals, long after being charged up.

If the effect of resistance is temporarily ignored, the opposition to the flow of an alternating current is the reactance (X).

Inductive reactance X_L is the reactance due to an inductance and:

$$X_L = 2\pi fL$$

(π is a mathematical constant which may be taken as 22/7 or 3.14 to two decimal places). X_L is in ohms when f is in hertz and L in henrys.

Similarly, capacitive reactance X_C is the reactance due to a capacitor and:

$$X_C = \frac{1}{2\pi fC}$$

X_C is in ohms when f is in hertz and C in farads. (Note that $2\pi f$ is often written as ω in mathematical formulae.)

Fig 2.6 shows the variation of reactance with frequency.

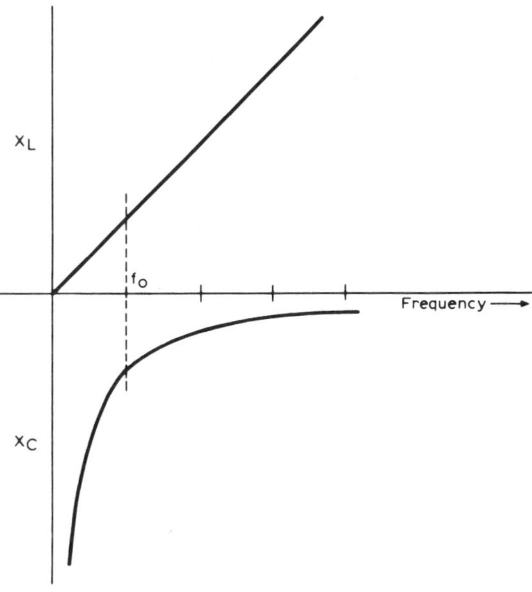

Fig 2.6. How the reactance of a capacitor and an inductor vary with frequency

When an alternating voltage V is applied to a resistance R the current which flows is exactly in step with the voltage. The voltage and current are said to be 'in phase'. The value of the current will be by Ohm's Law:

$$I = \frac{V}{R} \text{ amps}$$

When an alternating voltage, V, is applied to an inductor, L, (Fig 2.7) the current which flows will lag behind the voltage, the phase difference being 90°. The current flowing is given by:

$$I_L = \frac{V}{X_L}$$

where X_L is the inductive reactance, $2\pi fL$ ohms.

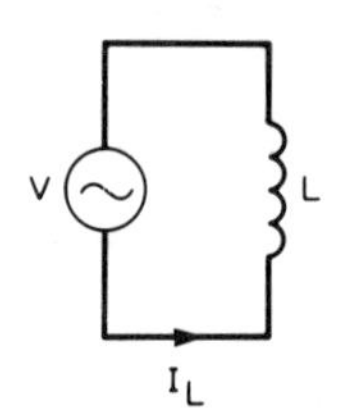

Fig 2.7. Alternating voltage applied to inductor L

When an alternating voltage, V, is applied to a capacitor, C, (Fig 2.8) the current leads the voltage by 90°. The current which flows is given by:

$$I_C = \frac{V}{X_C}$$

where X_C is the capacitive reactance, $1/(2\pi fC)$ ohms.

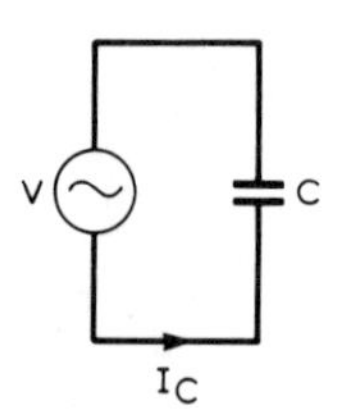

Fig 2.8. Alternating voltage applied to capacitor C

When an alternating voltage is applied to an inductor and a capacitor in series, the current flowing and the phase angle between the voltage and current will depend upon the effective reactance, X, of the circuit ie $(X_L - X_C)$ or $(X_C - X_L)$; thus the circuit will be inductive if X_L is greater than X_C or capacitive if X_C is greater than X_L.

If now there is a resistor, R, in series with the inductor and capacitor, the total opposition to the flow of an alternating current is known as the 'impedance', Z. This impedance is made up of resistance, R, and the effective reactance, X. Both R and X are measured in ohms, but they must not be added arithmetically. As a result of the 90° phase shift introduced by the inductance and capacitance, they must be added vectorially, ie by taking the square root of the sum of the squares of R and X, hence:

$$Z = \sqrt{(R^2 + X^2)}$$

Ohm's Law can now be applied and so the current flowing is:

$$I = \frac{V}{Z}$$

The following relationships should be noted:

Capacitors in series: $\frac{1}{C} = \frac{1}{C_1} + \frac{1}{C_2} + \frac{1}{C_3} + \ldots$

Capacitors in parallel: $C = C_1 + C_2 + C_3 + \ldots$

Inductors in series: $L = L_1 + L_2 + L_3 + \ldots$

Mutual inductance (which will be discussed later) is assumed to be zero.

Magnetism

Permanent magnets

The magnet and its properties of attracting a piece of iron by exerting a magnetic force on it and causing a compass needle to be deflected are well known.

Magnets made from certain types of steel and alloys of aluminium, nickel and titanium etc retain their magnetism more or less permanently. Such magnets find many uses in radio equipment such as in moving coil meters, headphones and loudspeakers.

Electromagnets

An electric current flowing through a straight wire creates a magnetic field, the lines of force of which are in a plane perpendicular to the wire and concentric with the wire.

The magnetic field surrounding such a straight wire is relatively weak, but a strong magnetic field can be produced by a current if, instead of a straight wire, a coil of wire or 'solenoid' is used; moreover, the field can be greatly strengthened if a piece of soft iron or other magnetic material (known as a 'core') is placed inside the coil.

The magnetic field produced by a solenoid is indeed similar to that produced by a bar magnet and it exhibits identical properties.

The extent by which the strength of the solenoid magnet is increased by the introduction of the core is called the 'permeability' of the core material. Permeability is really the ratio of the number of lines of force (or flux density) in the magnetic core to the flux density in a vacuum (ie no magnetic core). The difference between a vacuum and an air core is so small that it is ignored. As it is a ratio, strictly it should be referred to as 'relative permeability μ_r', but the word 'relative' is often omitted colloquially.

The strength of a magnetic field produced by a current is directly proportional to the current. It also depends on the

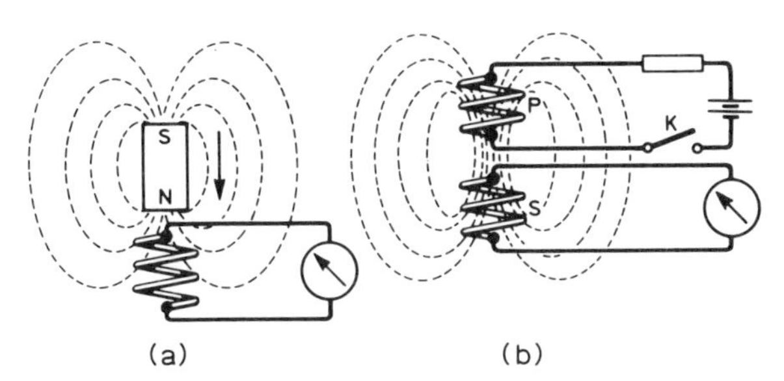

Fig 2.9. Electromagnetic induction. (a) Relative movement of magnet and coil causes a voltage to be induced in coil; (b) when current in one of a pair of coupled coils changes in value, a voltage is induced in the second coil

number of turns of wire, the area of the coil and the permeability of the core.

Electromagnetic induction

If a bar magnet is plunged into a solenoid, as indicated in Fig 2.9(a), the moving coil microammeter connected across the coil shows a deflection. The explanation of this phenomenon, known as 'electromagnetic induction', is that the movement of the magnet's lines of force past the turns of the coil causes a voltage to be induced in the coil which in turn causes a current to flow through the meter. The magnitude of the effect depends on the strength and speed of movement of the magnet and the size of the coil. Withdrawal of the magnet causes a reversal of the current. No current flows unless the lines of force are moving relative to the coil. The same effect is obtained if a coil of wire is arranged to move relative to a fixed magnetic field. Dynamos and generators depend for their operation on the principle of electromagnetic induction.

Suppose a pair of coils of wire is arranged as shown in Fig 2.9(b). When the switch K is open there is no magnetic field from the coil P linking the turns of the coil S, and the current through S is zero. Closing K causes a current in the coil P which produces a magnetic field. This field, as it builds up from zero, induces a voltage in S and causes a current to flow through the meter for a short time until the field due to P has reached a steady value, when the current through S falls to zero again. The effect is only momentary and is completed in a fraction of a second. The change in current in the circuit P is said to have 'induced' a voltage in the circuit S. The fact that a changing current in one circuit can induce a voltage in another circuit is the principle underlying the operation of transformers.

Self-inductance

If a steady current is flowing through a coil, there is a steady magnetic field due to that current. A current change tends to alter the strength of the field, which in turn induces in the coil a voltage (back EMF) tending to oppose the change being made. A negative sign is generally used before a back EMF to indicate that it is contrary to the supply voltage. This process is called 'self-induction'. A coil is said to have 'self-inductance', usually abbreviated to 'inductance'. It has a value of one 'henry' (H) if, when the current through the coil changes at a rate of 1A/s, the voltage appearing across its terminals is 1V. Inductance is usually denoted by the symbol L in formulae. As the inductance values used in radio equipment may be only a fraction of a henry, the units 'millihenry' (mH) and 'microhenry' (μH) (0.001 and 0.000001H respectively) are commonly used.

The inductance of a coil varies as the square of the number of turns, the cross-sectional area, and the permeability of the core, and inversely as the length of the magnetic path.

$$L \propto \frac{\mu_r A T^2}{l}$$

Mutual inductance

A changing current in one circuit can induce a voltage in a second circuit (see Fig 2.9(b)). The strength of the voltage induced in the second circuit S depends on the closeness or 'tightness' of the magnetic coupling between the circuits; for example, if both coils are wound together on an iron core, practically all the lines of force or magnetic flux from the first circuit link with the turns of the second circuit. Such coils are said to be 'tightly coupled' whereas if the coils are both air-cored and spaced some distance apart they are 'loosely coupled'.

The mutual inductance between two coils is measured in henrys, and two coils are said to have a 'mutual inductance' of 1H if when the current in the primary coil changes at a rate of 1A/s the voltage across the secondary is 1V. Mutual inductance is denoted in formulae by the symbol M.

The mutual inductance between two coils may be measured by joining the coils in series, first so that the sense of their windings is the same, and then so that they are reversed. The total inductance is then measured in each case.

If L_a and L_b are the total measured inductances, L_1 and L_2 are the separate inductances of the two coils and M is the mutual inductance, then:

$$L_a = L_1 + L_2 + 2M$$

$$L_b = L_1 + L_2 - 2M$$

$$L_a - L_b = 4M$$

$$M = \frac{L_a - L_b}{4}$$

The mutual inductance is therefore equal to one quarter of the difference between the series-aiding and series-opposing readings.

Inductors used in radio equipment

An inductor consists of a number of turns of wire. However, within the framework of this simple definition there is an extremely wide range of inductance values and types of construction. For example, an inductor required as a tuning coil at VHF might have an inductance of 0.5μH; this would probably be one or two turns of 2mm wire and self-supporting. At the other extreme, a smoothing choke in a power unit would have an inductance of, say, 30H and consist of between 1000 and 2000 turns wound on a paxolin bobbin with a laminated iron core. If this choke were designed to carry a current of 500mA, it might weigh about 6kg and occupy a 15cm cube in volume. Specialist low-frequency applications may require inductance values up to 500H.

The form of construction depends basically on the inductance value required. The number of turns of wire necessary to give the inductance depends on the permeability of its core. Air has a permeability of 1 but there are magnetic materials which have a very much higher permeability. Hence to achieve a reasonably high inductance without having to wind many thousands of turns of wire on an 'air' core, many fewer turns are wound on a magnetic core appropriate to the particular frequency involved.

The commonest magnetic core is made up of laminations, normally 0.3mm thick of silicon iron. Laminations are available in many shapes and sizes and are insulated on one side so that when they are assembled in a core they are insulated from each other. This reduces the power loss due to eddy currents induced in the core. The commonest shape of lamination is a pair, one being T-shaped and the other U-shaped, so that they fit together when assembled into the paxolin bobbin which carries the winding.

This type of core is standard for low-frequency chokes and power transformers. Thinner laminations of different types of iron are available for use at frequencies in the audio range.

Radio-frequency coils are usually air-cored, either self-supporting or wound on low-loss plastic or ceramic formers. The number of turns involved is often quite small, with inductance values up to about 20µH.

Larger values of inductance up to 1 or 2mH at frequencies from about 100kHz upwards may be wound on dust iron or ferrite cores. A dust iron core is a core of very finely divided iron alloy moulded in an insulating medium. Being moulded, different shapes and sizes can be made cheaply. Often a brass-threaded rod is moulded into a small cylindrical core. The position of the core within the coil can then be adjusted to vary the inductance value to tune the coil to a specific frequency. This is known as 'slug tuning'.

The modern ferrite cores are non-ferrous materials of high resistivity, and therefore low eddy current loss.

Moulded cores are often in two similar halves as a 'pot' core. The winding is put on a small plastic bobbin which goes inside the two halves and so is surrounded by the ferrite material. Ferrite materials are also moulded in the form of a ring or 'toroid' (a 'toroidal' core). Such cores ensure that the magnetic flux is nearly all contained within the windings, ie there is practically no stray field.

Capacitors used in radio equipment

'Capacitance' may be defined as the ability of a conductor to store an electric charge. A device in which this effect is enhanced is called a 'capacitor'. In its simplest form, the capacitor consists of two parallel plates as shown in Fig 2.10. The material between the plates is known as the 'dielectric'; in this case the dielectric is air.

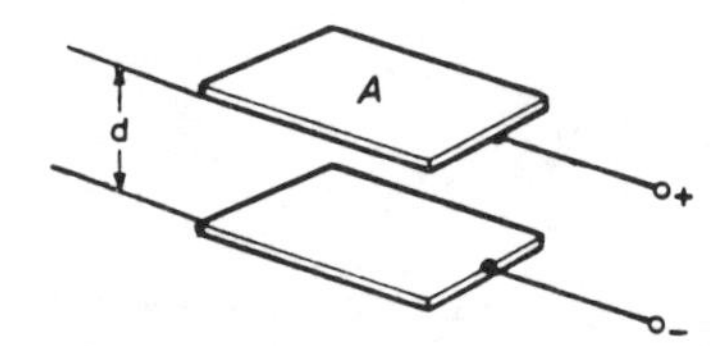

Fig 2.10. Parallel-plate capacitor. The capacitance is proportional to area *A* and inversely proportional to spacing *d*

The capacitance of such a capacitor is proportional to the area of the plates and inversely proportional to the distance between them. It also depends upon a property of the dielectric known as the 'permittivity' (the dielectric constant). As in the case of permeability, referred to earlier, permittivity is a quantity referred to the vacuum (or air) state. It is hence a ratio or relative property.

Typical relative permittivities are:

air	1
paper	2 approximately
polythene, rubber etc	2.3 "
mica	5 "
ceramics	10 and upwards

The dielectric material determines the maximum frequency at which a capacitor may be used, eg paper up to 20MHz or so, some ceramics and plastics (polythene or polystyrene) up to 150 to 200MHz, and mica even higher.

The capacitor exists in many forms and is classified by the material used as the dielectric. The range of capacitance commonly available is from 1pF to at least 68,000µF. Capacitors have the property of being able to store a charge of electricity and must be capable of withstanding a voltage difference between the plates. Thus the larger the plates and/or the smaller their separation, the greater is the charge that the capacitor holds for a given voltage across its plates. Ultimately, if the voltage between the plates is too high the capacitor will flash over or break down. The safe working voltage is therefore an important property of a capacitor. Capacitors are made with working voltages from 6V to many thousands of volts.

It is clear from its form of construction that a capacitor presents an open circuit to a direct current but it appears to pass an alternating current because of the build-up and decay of charge on one plate and then the other as the direction of flow of the alternating current changes.

Mica capacitors are normally made up to a value of 0.01µF and may be used up to very high frequencies (VHF). They consist of a stack of plates interleaved with layers of mica, which are clamped together, dipped in wax or potted in resin. In the silvered-mica type, the plates are made by spraying a very thin layer of silver on to the mica dielectric.

Synthetic materials such as polystyrene, polytetrafluoroethylene (PTFE) and polycarbonate in the form of thin film have now generally replaced paper as a dielectric as their loss as a dielectric is lower than that of the latter. Hence, they can be used at higher frequencies than the normal limit of about 20MHz for paper.

Ceramic capacitors are made by spraying silver 'plates' on to both sides of a ceramic cup, disc or tube. The ceramic used has high permittivity so that a high capacitance is obtained in a small volume. They are made in relatively low working voltages and are most commonly used as bypass capacitors at VHF.

The feedthrough capacitor is a form of ceramic capacitor in which one plate is a threaded bush on the outside of a ceramic tube (the 'dielectric'), and the other plate a stiff wire through the centre of the tube. This type is used for feeding through power supplies into a screened box; thus it combines a feedthrough insulator with a bypass capacitor.

Electrolytic capacitors have plates of aluminium or tantalum foil with a semi-liquid conducting compound, often in the form of impregnated paper, between them. The dielectric is a very thin insulating layer which is formed by electrolytic action on one of the foils when a DC polarising voltage is applied to the capacitor. As the dielectric is very thin, very high values of capacitance can be put into a small space. The capacitance value can be 68,000μF or more. There is a small leakage current through an electrolytic capacitor and it must be emphasised that generally they are polarised, ie one terminal is positive and the other is negative. Although electrolytic capacitors can withstand a small ripple (alternating) current, this polarity must be strictly observed otherwise the capacitor may explode.

The capacitor has many uses:

(a) as part of a tuned circuit;
(b) as a coupling capacitor between two stages in an RF or AF amplifier, it passes AC but holds off the direct voltage supply to the first stage from the input to the following stage;
(c) a capacitor of the appropriate value provides a low-impedance path at a particular frequency or range of frequencies and hence is used to bypass to earth unwanted AF or RF voltages which may occur on the direct voltage supply line (this is known as 'decoupling'). The smoothing capacitors in a power supply perform an identical function by bypassing harmonics of the supply frequency to earth.

The decibel notation

The need to compare voltage levels or power levels at different points in a circuit or at different frequencies very often arises in radio engineering.

The most realistic way to do this is by means of the decibel notation which is based on logarithms as the following example shows.

Consider the statement 'the power level has increased by 1W.' What does this mean? Obviously an increase in power from 0.25W to 1.25W is vastly different from an increase from 10W to 11W or from 100W to 101W, yet in each case the power level has increased by 1W.

Table 2.3. Ratios of power and voltage in terms of decibels

dB	Power ratio	Voltage ratio	dB	Power ratio	Voltage ratio
1	1.26	1.12	15	31.6	5.62
2	1.58	1.26	20	100	10
3	2.00	1.41	30	1000	31.6
4	2.51	1.58	40	10^4	10^2
5	3.16	1.78	50	10^5	316
6	3.98	2.00	60	10^6	10^3
7	5.01	2.24	70	10^7	3160
8	6.31	2.51	80	10^8	10^4
9	7.94	2.82	90	10^9	31600
10	10.00	3.16	100	10^{10}	10^5

The effect of a 1W increase in power in each case may be compared with the use of the decibel notation. The difference between a power level W_1 and a power level W_2 when expressed in decibels is:

$$\text{Number of decibels} = 10 \log_{10} \frac{W_2}{W_1}$$

In the above examples

(a) 0.25W to 1.25W

$$\text{Increase in decibels} = 10 \log_{10} \frac{1.25}{0.25} = 7\text{dB}$$

(b) 10W to 11W

$$\text{Increase in decibels} = 10 \log_{10} \frac{11}{10} = 0.4\text{dB}$$

(c) 100W to 101W

$$\text{Increase in decibels} = 10 \log_{10} \frac{101}{100} = 0.04\text{dB}$$

The advantage of this notation is therefore obvious. Voltages may also be compared in this way, since:

$$W = \frac{V^2}{R}$$

$$\frac{W_2}{W_1} = \frac{V_2^2}{R_2} \div \frac{V_1^2}{R_1}$$

$$= \frac{V_2^2}{R_2} \times \frac{R_1}{V_1^2}$$

If, and only if, $R_1 = R_2$ then:

$$\frac{W_2}{W_1} = \frac{V_2^2}{V_1^2}$$

$$\text{hence number of decibels} = 10 \log_{10} \frac{V_2^2}{V_1^2}$$

$$= 20 \log_{10} \frac{V_2}{V_1}$$

Other ratios may be calculated easily from the above. If two decibel figures are added, the corresponding power or voltage ratios must be multiplied, eg

$$45\text{dB} = 40\text{dB} + 5\text{dB} = 100 \times 1.78 \text{ (voltage ratios)}$$
$$= 178$$

The decibel notation is a way of expressing a ratio, therefore it can only be used to express a magnitude when a reference level is defined. This is often done as follows.

The unit 'dBW' specifies a power level which is so many decibels above one watt, ie the reference level. Thus +20dBW is a power level which is 20dB above one watt, ie 100 watts. The + sign here is quite often omitted. (Note 0dBW is zero dB above 1 watt, ie it is 1 watt!)

This is the method now used to express transmitter output power in the UK amateur radio licence.

Powers less than 1 watt are also expressed in this way, eg –20dBW means 20dB down on 1 watt, that is 10mW (the negative sign here is never omitted).

Similarly, 20dBmW is 20dB up on 1mW or 100mW. And –20dBmW is 20dB below 1mW, that is 10μW.

Other common examples of the use of the decibel are:

(a) The variation in gain of an audio amplifier at different frequencies may be expressed as so many decibels above or below the gain at, say, 1000Hz.
(b) The characteristic of a bandpass filter is specified by its shape factor which is the ratio of its bandwidths at –60dB and –6dB.
(c) The variation in level of the selectivity or response curve of a tuned circuit or receiver is expressed in decibels (see Fig 5.6).
(d) The AGC characteristic of a receiver is a graph of the variation in output of the receiver (in decibels) plotted against increasing signal input (also in decibels).

Other reference levels used are 'dBV' and 'dBmV'. 'dBd' and 'dBi' are used to compare the gain of an antenna with reference to a dipole or an isotropic radiator respectively.

Note: 'dBmW' is commonly abbreviated to 'dBm'.

Acoustic power level (ie loudness) is also specified in decibels, a standard reference level being assumed.

Tuned circuits

The parallel and series connections of an inductor and a capacitor to form a tuned circuit are shown in Figs 2.11(a) and 2.11(b) respectively.

At one particular frequency, the numerical values of the reactance of the inductor and the reactance of the capacitor will be equal (see Fig 2.6), that is:

$$X_L = X_C$$

or

$$2\pi f_r L = \frac{1}{2\pi f_r C}$$

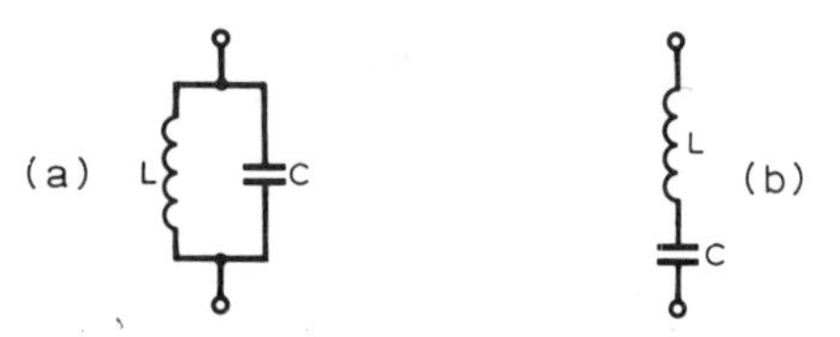

Fig 2.11. Arrangement of capacitor and inductor to form tuned circuit: (a) parallel-tuned circuit; (b) series-tuned circuit

This can be simplified to give:

$$f_r = \frac{1}{2\pi\sqrt{LC}}$$

(f_r is in hertz when L is henrys and C in farads). f_r is called the 'resonant frequency' of the tuned circuit.

There are inevitably losses associated with all tuned circuits because neither the inductor nor the capacitor is perfect; these losses are always assumed to be resistive and are shown, when required, as a resistor R in series with the inductor.

Consider the circuit of Fig 2.12. It can be shown that the impedance at resonance of this circuit is L/CR ohms. Hence by Ohm's Law:

$$V = I \times \frac{L}{CR}$$

and

$$I_C = \frac{V}{X_C}$$

$$= I \times \frac{L}{CR} \times 2\pi fC$$

(It is convenient here to use ω rather than $2\pi f$.)

hence

$$I_C = I \times \frac{\omega L}{R}$$

Similarly it can be shown that the current which flows through the inductor is:

$$I_L = I \times \frac{\omega L}{R}$$

Thus the currents flowing through the capacitor and the inductor I_C and I_L are both greater than the input current I by a factor of $\omega L/R$. This factor is called the 'magnification factor' of the tuned circuit. It is generally just known as the 'Q' and may be quite large, thus:

$$Q = \frac{\omega L}{R}$$

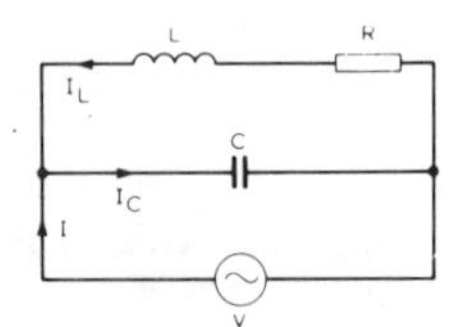

Fig 2.12. Current and voltage relationships in a parallel-tuned circuit

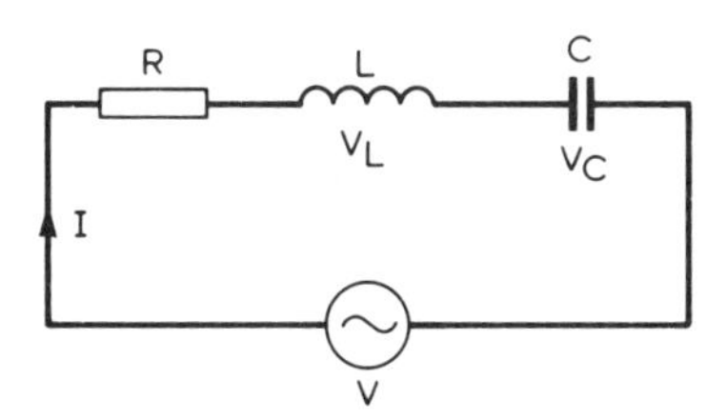

Fig 2.13. Current and voltage relationships in a series-tuned circuit

The current which flows through L and C is known as the 'circulating current'. This, as shown above, can be quite large although the current I taken from the supply V may be small.

As a capacitor generally has quite low losses when correctly used, the Q of a tuned circuit is determined by the inductor. The range of Q obtainable is roughly 100 to 400 depending on the type and form of the inductor. The Q may be reduced by the load placed upon the tuned circuit by the circuit which follows it.

The currents flowing through the inductor and capacitor vary with the frequency as they depend upon the term

$$\frac{\omega L}{R} \quad \left(\text{ie } \frac{2\pi fL}{R}\right)$$

If the ratio of I_L and I is plotted against frequencies above and below the resonant frequency, a curve similar in shape to Fig 2.15 is obtained. This curve is called a 'response curve' because it indicates how the tuned circuit responds to different frequencies; it is also known as a 'resonance curve'. The shape of this curve is determined by the Q – the greater the Q, the higher and narrower the curve becomes.

The above discussion also applies to the series-tuned circuit (see Fig 2.13) in which the impedance is at a minimum at resonance and is equal to R. The voltages across the inductor and capacitor, not the current flowing in each, are calculated as before and the increase in the voltages is equal to Q. If in this case, the ratio V_L/V_C is plotted, a similar resonance curve is obtained.

The value of the impedance of the parallel-tuned circuit at resonance is known as the 'dynamic resistance'. This is a fictitious resistance and exists for alternating currents of the resonant frequency. Its symbol is R_D.

The dynamic resistance can be expressed in terms of Q as follows:

$$R_D = \frac{L}{CR}, \quad \text{since } Q = \frac{\omega L}{R}, \quad R_D = \frac{Q}{\omega C}$$

A good-quality tuned circuit will have a R_D of about 50,000Ω. The DC resistance is usually very low.

The parallel-tuned circuit is by far the most commonly used one.

The series-tuned circuit, having minimum impedance at resonance, accepts maximum current at resonance and is sometimes known as an 'acceptor circuit' (see Fig 2.14(a)).

In a parallel-tuned circuit, the impedance is at maximum at resonance. The parallel-tuned circuit is often called a 'rejector circuit' because, having a high impedance at resonance, it rejects current at this frequency (see Fig 2.14(b)).

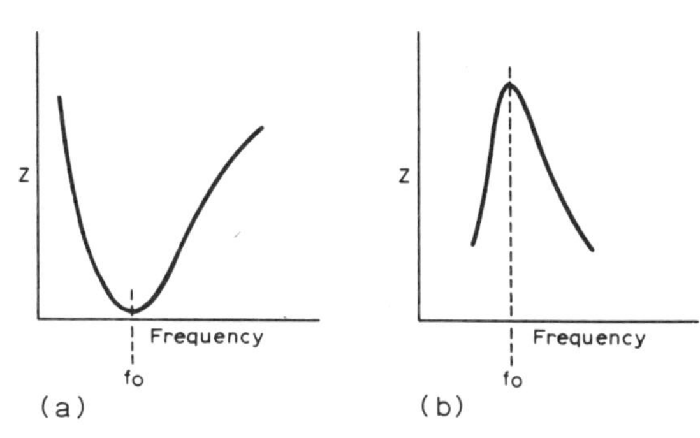

Fig 2.14. Impedance/frequency characteristics of (a) series-tuned circuit; (b) parallel-tuned circuit

L/C ratio

The resonant frequency of a tuned circuit is fixed by the product of inductance and capacitance, ie $L \times C$. There is hence an infinite number of values of L and C which will tune to a given frequency.

The choice of the ratio of L and C is determined by practical considerations according to the particular use.

For instance, in receiver tuned circuits, in order to achieve a high value of dynamic resistance (required for high gain) the L/C ratio is high. The limitation here is that there is always a minimum value of capacitance inherent in any circuit (this is the 'stray capacitance').

In some applications, it may be required to swamp completely the stray circuit capacitance, in which case the L/C ratio is low.

The optimum choice of L/C ratio is sometimes difficult. As a compromise, it is convenient to assume that the value of C is 1.5pF per metre of wavelength; thus to tune to 30MHz (ie 10m) a capacitance of about 15pF would be reasonable.

Resonance curves and selectivity

The ability of a tuned circuit to differentiate between a wanted frequency and an adjacent unwanted frequency is its 'selectivity'. This depends upon its Q; good selectivity requires high Q.

Fig 2.15 is a typical response curve (or selectivity curve) of a tuned circuit. The width of this curve, ie $f_h - f_l$, is known as the 'bandwidth' of the tuned circuit. Bandwidth can be defined at various levels, eg at the level where the response has fallen to $1/\sqrt{2}$ or 0.707 (70.7%) of the maximum response (this is also known as the 'half-power point' or –3dB level). The ratio of the widths at –60dB and –6dB is known as the 'shape factor'.

The relationship between bandwidth and Q is:

$$Q = \frac{f_r}{f_h - f_l} = \frac{f_r}{2\Delta f}$$

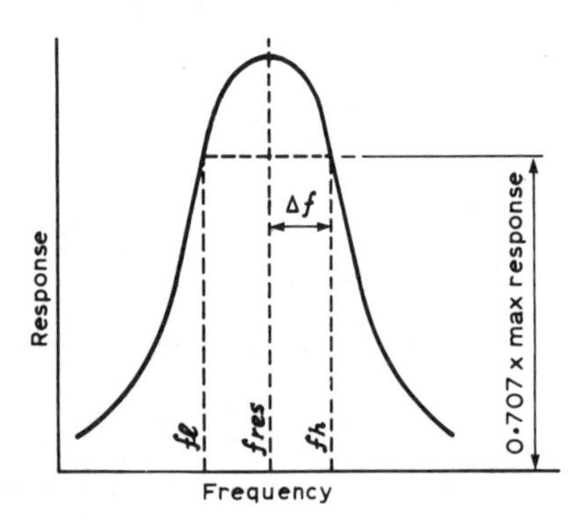

Fig 2.15. Resonance curve of parallel-tuned circuit

where f_r is the resonant frequency and $2\Delta f$ is $f_h - f_l$. From this it is seen that the lower f is, the lower Δf is, ie the bandwidth is less and selectivity is greater for a given value of Q. Similarly, the bandwidth is inversely proportional to Q; ie the higher the Q, the smaller the bandwidth.

The effect of several tuned circuits in cascade, eg in successive stages of an amplifier, is to reduce the overall bandwidth, ie increase the selectivity. This is particularly apparent in the region known as the 'skirts' of the selectivity curve where the bandwidth is greatest. In the majority of cases, the design aim is to increase the effective overall selectivity, but in some instances it is necessary to decrease the selectivity or increase the bandwidth. This is achieved by a 'damping resistor' connected across the tuned circuit; the lower the resistor, the greater its effect.

Coupled circuits

Pairs of coupled tuned circuits are often used in receivers and transmitters. The effect of varying the degree of coupling between two parallel tuned circuits resonant at the same frequency is shown in Fig 2.16.

When the coupling is loose, the response from one circuit to the other is as curve I. As the coupling is increased to what is known as 'critical coupling', the output at resonance increases to curve II; here the mutual coupling between the coils is $1/Q$ of the inductance of either coil. Further increase (tight coupling) results in the formation of the double-humped characteristic shown in curve III, where the output at resonance has decreased.

Two tuned circuits are often mounted in a screening can, the coils generally being wound the necessary distance apart on the same former to give the required coupling. The coupling is then said to be 'fixed'.

Transformers

The principle of electromagnetic induction is illustrated in Fig 2.9(b). This is the basis of the transformer in which an alternating current flowing in the winding P creates an alternating magnetic field which cuts the turns of the winding S. It is therefore obvious that transformers can only operate on an alternating supply!

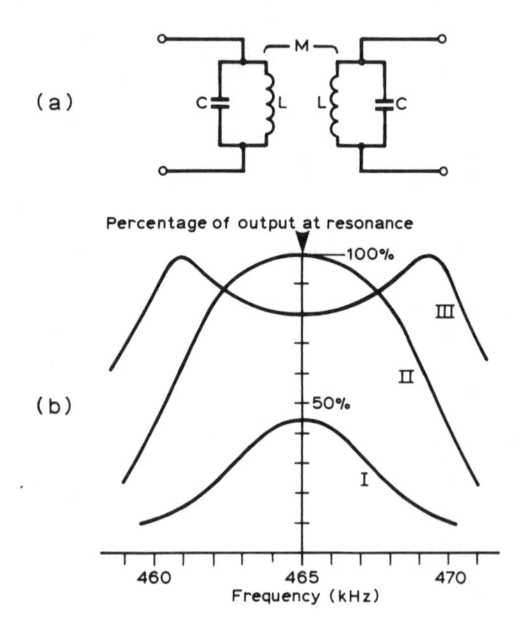

Fig 2.16. Inductively-coupled tuned circuits. Curves shown at (b) represent various frequency-response characteristics of coupled circuit shown at (a) for different degrees of coupling

Transformers perform many vital functions in electrical and radio engineering; for example, the transfer of electrical energy from one circuit to another and, implied in the latter, the transformation of an alternating voltage upwards or downwards.

If in a transformer the number of turns on the primary winding is n_p, the number of turns on the secondary winding is n_s, the voltage across the primary is V_p and the voltage across the secondary is V_s, then:

$$V_s = \frac{n_s}{n_p} \times V_s \quad \text{or} \quad \frac{V_s}{V_p} = \frac{n_s}{n_p}$$

The term n_s/n_p is called the 'turns ratio', which may be less than or greater than unity; thus the transformer may 'step down' or 'step up'. The relationship between the currents in the primary and secondary windings is similarly:

$$I_p = \frac{n_s}{n_p} \times I_s \quad \text{or} \quad \frac{I_p}{I_s} = \frac{n_s}{n_p}$$

If primary impedance is Z_p and secondary impedance is Z_s, then:

$$Z_p = \left(\frac{n_p}{n_s}\right)^2 Z_s$$

$$\text{or} \quad \frac{Z_p}{Z_s} = \left(\frac{n_p}{n_s}\right)^2$$

This is particularly important; note that the impedance ratio is equal to the turns ratio squared whereas the voltage ratio equals the turns ratio itself.

Power transformers are normally wound on a bakelite bobbin through which a laminated silicon iron core is assembled, as in the case of the iron-cored inductor.

Filters

Filters, or to give them their full name 'wave filters', are passive networks of capacitors and inductors which exhibit certain characteristics as the input frequency is varied.

The filters of most interest in amateur radio are:

(a) *Low-pass filters.* A low-pass filter passes all frequencies below a specified frequency but attenuates frequencies above it.
(b) *High-pass filters.* A high-pass filter passes all frequencies above a specified frequency but attenuates frequencies below it.

The specified frequency referred to is the 'cut-off frequency' (f_C). The configuration of the simplest form ('single section') of each filter is shown in Fig 2.17 which also shows the general shape of the characteristics.

Two or three (or occasionally more) single-section filters may be connected in cascade to increase the rate of the fall-off of the response in the stop band. Further improvement may be achieved by connecting a series arrangement of inductance and capacitance ('half sections') across the input and output of the filter.

Although not strictly a wave-filter, a 'T-notch filter' has uses in amateur radio. This provides a tuneable and very sharp 'null', ie a large attenuation over a narrow frequency band.

Filters have a number of important applications in amateur radio. Low-pass filters are used to attenuate unwanted frequencies in the output of the HF bands transmitter (transmitter-antenna matching unit connection). Another use is to limit the audio bandwidth of a telephony transmitter to the minimum necessary for intelligible communication.

The high-pass filter may be used in the antenna downlead (coaxial cable) of a television receiver in order to attenuate unwanted frequencies.

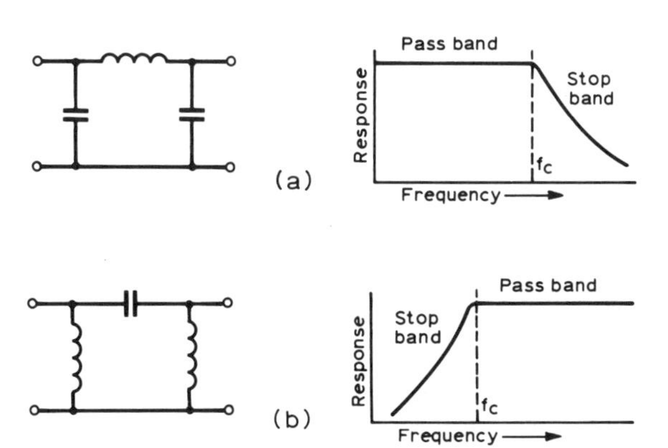

Fig 2.17. (a) Single-section low-pass filter configuration and its response curve; (b) single-section high-pass filter configuration and its response curve

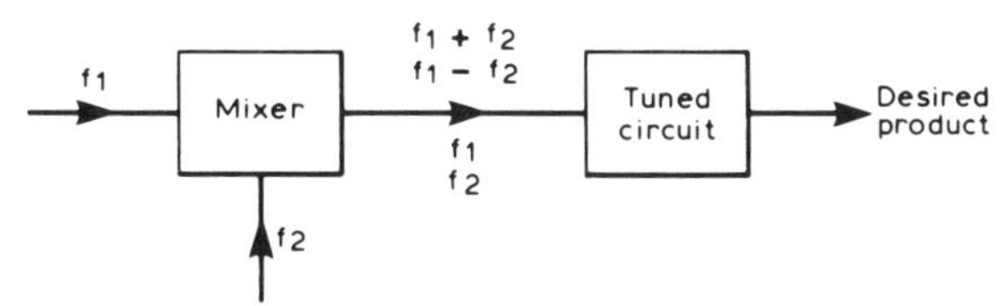

Fig 2.18. Process of frequency mixing

Mixing

The 'mixing' or 'heterodyne' process is one in which two signals are mixed to produce two new signals, one of which is equal in frequency to the sum of the original frequencies and the other equal to the difference between them. The undesired product and the two original frequencies are rejected by some form of filter which is generally a tuned circuit. This process is also called 'frequency changing', 'frequency conversion' or 'frequency translation' and is shown in Fig 2.18.

The mixer has many uses in communication engineering. In particular, it is the basis of the superheterodyne receiver and the single-sideband (SSB) transmitter.

Owing to non-optimum characteristics in the mixer element, spurious products harmonically related to the two original frequencies (f_1 and f_2) are often produced. The suppression of these is particularly important in an SSB transmitter. They may be reduced by careful mechanical layout of the circuit and the use of a push-pull, ie balanced, circuit.

Tolerance and effect of temperature on components

These are important aspects of all electronic components, particularly resistors and capacitors.

Tolerance

The tolerance on the value of a component is a measure of how accurate its value is. For example, a resistor of nominal value 10,000Ω, and a tolerance of ±10%, may have an actual resistance between 9000Ω (ie 10,000 – 10%) and 11,000Ω (ie 10,000 + 10%).

The usual tolerances on the values of resistors and capacitors are ±20, ±10, ±5, ±2 and ±1%.

Temperature effects

Generally the value of a component increases as the temperature increases. A constant known as the 'temperature coefficient' is a measure of this increase, and thus (for most items) the temperature coefficient is positive.

There are a few exceptions, eg ceramic capacitors having a negative temperature coefficient can be made. Carbon and semiconductors have a negative coefficient.

Preferred values

The three common ranges of basic 'preferred values' for components are listed in Table 2.4. The preferred values

Table 2.4. Preferred values

20% tolerance	10% tolerance	5% tolerance
10	10	10
–	–	11
–	12	12
–	–	13
15	15	15
–	–	16
–	18	18
–	–	20
22	22	22
–	–	24
–	27	27
–	–	30
33	33	33
–	–	36
–	39	39
–	–	43
47	47	47
–	–	51
–	56	56
–	–	62
68	68	68
–	–	75
–	82	82
–	–	91

which are normally available are multiples of the basic values between 1 and $10^7\Omega$, for example 4.7Ω, 47Ω, 470Ω, 4.7kΩ, 47kΩ, 470kΩ and 4.7MΩ (20% tolerance) resistors. The manufacture of preferred values only is general for resistors, but less so for capacitors, particularly above 1μF.

Quartz crystals

A quartz crystal is a very thin slice of quartz which has been cut from a large, naturally occurring crystal of quartz. Quartz exhibits the 'piezo-electric' effect, ie a mechanical strain applied to a suitably cut piece of quartz causes an electric stress to be set up between opposite faces of the piece. Conversely an electric stress applied between opposite faces of the piece causes a mechanical deformation. The frequency of resonance of this mechanical deformation depends upon the dimensions of the slice and the mode of vibration; other properties (eg temperature coefficient) depend upon the orientation of the slice with reference to an axis of the natural crystal.

Crystals can operate up to a frequency of about 22MHz in the fundamental mode and up to about 200MHz in the appropriate circuit in what is known as the 'overtone' mode. Overtone operation occurs at frequencies close to the odd multiples of the fundamental frequency.

Crystals for frequencies below about 1MHz are generally in the form of a bar rather than a thin slice; at 20kHz, for example, this bar is about 70mm long.

Connections to the modern crystal are made to electrodes of gold or silver which are deposited on opposite faces of the crystal. These connections also support the crystal, which is then hermetically sealed in an evacuated glass envelope or a small cold-welded metal container.

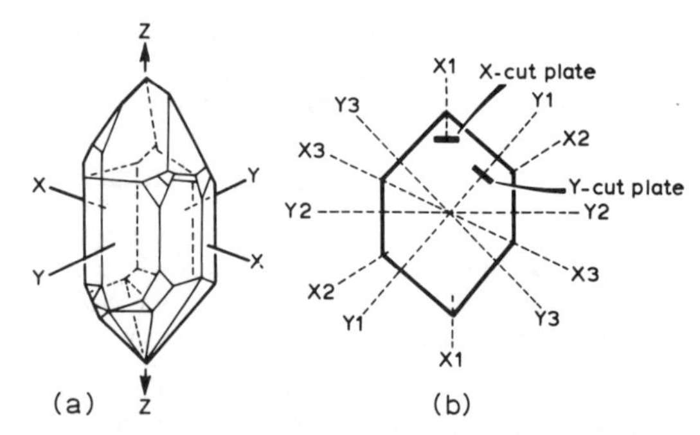

Fig 2.19. (a) Double-terminated natural quartz crystal indicating relationship between X, Y and Z axes. The crystal usually has one pyramidal termination and a rough end where it has broken from the parent rock. (b) Various X and Y axes as viewed along a direction parallel to the Z axis and examples of positions of X-cut and Y-cut plates

The frequency of a crystal can be changed or 'pulled' by a few kilohertz by variation of capacitance in parallel with it (a varactor diode or a variable capacitor).

The quartz crystal is equivalent to a very-high-*Q* tuned circuit and so can be subsituted for the tuned circuit in an oscillator (see Chapter 3). Crystals are also used in pairs in band-pass filters.

Screening

It is often necessary to restrict the magnetic field around an inductor to prevent coupling between that inductor and another one close to. This is achieved by enclosing the inductor in an earthed metal can, this being known as 'screening'. The can should be at least 1½ times the coil diameter from the coil, otherwise the *Q* of the latter will be degraded. Generally, it is not necessary to screen coils which are separated by at least three coil diameters or which are tuned to different frequencies.

Similarly it is often required to screen one part of a circuit from another, eg to prevent feedback from the output of a high-gain amplifier to its input, or to prevent an unwanted strong signal getting to the input of an amplifier.

Aluminium and preferably copper are satisfactory screening materials at radio frequencies, but a high-permeability alloy such as mu-metal is necessary at audio frequencies.

Calculations

A number of sample calculations involving resistance, capacitance, inductance, reactance and resonance follow.

The mathematics involved are explained in Appendix 2. The calculations required in the RAE are generally more simple than these and often the answers can be obtained by inspection or approximation.

Example 1

A current of 50mA flows through a resistor of 1.5kΩ. What is the voltage across the resistor?

By Ohm's Law

$$V = I \times R$$
$$V = \frac{50}{1000} \times 1500$$
$$V = 75\text{V}$$

Example 2

In a stage of a receiver, 12V are applied across a potential divider of 3300Ω and 2700Ω. What is the current through the resistors?

By Ohm's Law

$$I = \frac{V}{R} = \frac{12}{3300 + 2700}$$
$$I = 0.002\text{A} = 2\text{mA}$$

Example 3

Resistors of 33kΩ and 27kΩ are connected in series. What is the effective resistance?

$$R = R_1 + R_2$$
$$R = 33{,}000\Omega + 27{,}000\Omega$$
$$R = 60{,}000\Omega$$

Example 4

Resistors of 100Ω and 150Ω are connected in parallel. Find the effective resistance.

$$\frac{1}{R} = \frac{1}{R_1} + \frac{1}{R_2}$$
$$\frac{1}{R} = \frac{1}{100} + \frac{1}{150}$$
$$\frac{1}{R} = \frac{6}{600} + \frac{4}{600} = \frac{10}{600}$$

Therefore, by cross-multiplying

$$600 = 10R$$
$$R = 60\Omega$$

Alternatively, the value of two resistors in parallel can be found by dividing the product of their values by the sum.

$$\frac{150 \times 100}{250} = 60\Omega$$

Example 5

In the smoothing circuit of a power supply, capacitors of 8μF, 4μF and 2μF are connected in parallel. What is the effective capacitance?

$$\text{Effective capacitance} = 8 + 4 + 2 = 14\mu\text{F}$$

Example 6

Capacitors of 220pF, 470pF and 0.001μF are connected in parallel. What is the effective capacitance?

Before addition can be effected, the values must first of all be expressed either in picofarads or in microfarads. Since there are 1,000,000pF to 1μF,

$$0.001\mu\text{F} = 1{,}000{,}000 \times 0.001\text{pF} = 1000\text{pF}$$

Therefore the effective capacitance is

$$220 + 470 + 1000 = 1690\text{pF}$$

Alternatively,

$$220\text{pF} = \frac{220}{1{,}000{,}000} = 0.00022\mu\text{F}$$

and

$$470\text{pF} = \frac{470}{1{,}000{,}000} = 0.00047\mu\text{F}$$

so that the effective capacitance is

$$0.00022 + 0.00047 + 0.001 = 0.00169\mu\text{F}$$

Example 7

Two capacitors of 0.001μF and 0.0015μF respectively are connected in series. Find the effective capacitance.

$$C = \frac{C_1 \times C_2}{C_1 + C_2} = \frac{1000 \times 1500}{1000 + 1500}$$
$$\frac{1000 \times 1500}{2500} = 600\text{pF}$$

Example 8

Two inductors of 10 and 20μH are connected in series; two others of 30 and 40μH are also connected in series. What is the equivalent inductance if these series combinations are connected in parallel? Assume that there is no mutual induction.

The 10 and 20μH coils in series are equivalent to (10 + 20) = 30μH

The 30 and 40μH coils in series are equivalent to (30 + 40) = 70μH

These two equivalent inductances of 30μH and 70μH respectively are in parallel and therefore equivalent to one single inductance of

$$\frac{30 \times 70}{30 + 70} = 21\mu\text{H}$$

Example 9

What power is consumed by a transmitter taking 1.5A at 12V?

$$W = V \times I \text{ watts} = 12 \times 1.5 = 18\text{W}$$

Example 10

What is the input power of a transmitter stage running at 24V, 2.5A?

$$W = V \times I \text{ watts} = 24 \times 2.5 = 60\text{W}$$

Example 11

Find the power dissipated by a 15Ω resistor when it is passing 1.2A.

$$W = I^2 \times R = 1.2^2 \times 15 = 21.6\text{W}$$

Example 12

The current at the centre of a given λ/2 antenna is found to be 0.5A. If this antenna has a radiation resistance of 70Ω, find the radiated power.

$$W = I^2 \times R = (\tfrac{1}{2} \times \tfrac{1}{2}) \times 70$$

$$= \frac{70}{4} = 17.5\text{W}$$

Example 13

A transmitter output stage is running at 20V, 3A. It is found to produce a current of 0.9A RMS in a load resistance of 50Ω. Find (a) the input power, (b) the output power, and (c) the efficiency of the stage.

(a) input power $= W =$ DC volts × DC amperes
$= 20 \times 3 = 60\text{W}$

(b) output power $= W =$ (load current)^2 × load resistance
$= 0.9 \times 0.9 \times 50 = 40.5\text{W}$

(c) efficiency $= \eta = \dfrac{\text{output power}}{\text{input power}}$

$$\eta = \frac{40.5}{60} = 0.675$$

or $\eta = 0.675 \times 100 = 67.5\%$

Example 14

What is the reactance of a 15H smoothing choke at a frequency of (a) 50Hz and (b) 400Hz?

(a) $X_L = 2\pi fL = 2\pi \times 50 \times 15$
$X_L = 4700$ (approximately)

(b) At eight times the frequency, X is eight times as great, ie 37,600Ω.

Example 15

A medium-wave coil has an inductance of 150μH. Find the reactance at a frequency of 500kHz.

$$X_L = 2\pi fL = 2\pi \times 500 \times 1000 \times \frac{150}{1{,}000{,}000}$$

$$= 470\Omega \text{ (approximately)}$$

Note how the frequency in kilohertz must be multiplied by 1000 to bring it to hertz, and how the inductance in microhenrys must be divided by 1,000,000 to bring it to henrys.

Example 16

What is the reactance of a 2μF smoothing capacitor at a frequency of (a) 50Hz, and (b) 400Hz?

(a) $$X_C = \frac{1}{2\pi fC}$$

where f is the frequency and C the capacitance in farads.

$$X_C = \frac{1}{2\pi \times 50 \times \dfrac{2}{1{,}000{,}000}}$$

$$X_C = \frac{1{,}000{,}000}{2\pi \times 100} = \frac{100{,}000 \times 3.2}{200}$$

$$X_C = 1600 \text{ (taking } 10/\pi = 3.2)$$

(b) At eight times the frequency, X is one-eighth of the above value, ie 200Ω.

Example 17

A coil has a resistance of 3Ω and a reactance of 4Ω. Find the impedance.

$$Z = \sqrt{(R^2 + X^2)}$$
$$= \sqrt{(3^2 + 4^2)}$$
$$= \sqrt{(9 + 16)} = \sqrt{25}$$
$$Z = 5\Omega$$

Example 18

Given a series circuit with a resistance of 60Ω and a capacitor with a reactance (at the working frequency) of 80Ω, find the impedance of the circuit.

$$Z = \sqrt{(R^2 + X^2)}$$
$$= \sqrt{(60^2 + 80^2)}$$
$$= \sqrt{(3600 + 6400)}$$
$$= \sqrt{10{,}000}$$
$$Z = 100\Omega$$

Example 19

Suppose the coil of Example 17 were connected (a) across 15V DC and, (b) across 15V AC (of frequency at which the reactance was 4Ω). Find the current in each case.

(a) At DC the reactance is zero and only the resistance opposes the passage of current. By Ohm's Law

$$I = \frac{V}{R} = \frac{15}{3} = 5\text{A}$$

(b) At AC Ohm's Law may still be used, provided Z, the impedance, is used in place of R, the resistance.

$$I = \frac{V}{Z} = \frac{15}{5} = 3\text{A}$$

Example 20

In the series circuit of Example 18, suppose the circuit were connected (a) across 240V DC and (b) across 240V AC (of frequency at which the reactance was 70Ω). Find the current in each case.

(a) A capacitor blocks the passage of direct current, therefore the current is zero amperes.

(b) Ohm's Law still holds, provided Z (the impedance) is used in place of R (the resistance).

$$Z = \sqrt{(60^2 + 70^2)} = 92.2$$

$$I = \frac{E}{Z} = \frac{240}{92.2} = 2.6\text{A}$$

Example 21

Find the capacitance required to resonate a 10H choke to 500Hz.

For resonance

$$2\pi fL = \frac{1}{2\pi fC}$$

hence

$$C = \frac{1}{4\pi^2 f^2 L} \text{ farads}$$

$$= \frac{1{,}000{,}000}{4\pi^2 f^2 L} \text{ microfarads}$$

and inserting the given values

$$C = \frac{1{,}000{,}000}{4\pi^2 \times 500 \times 500 \times 10}$$

$$\approx \frac{10}{100 \times 10} = 0.01\mu\text{F}$$

Example 22

A coil of 100μH inductance is tuned by a capacitance of 250pF. Find the resonant frequency.

For resonance

$$2\pi fL = \frac{1}{2\pi fC}$$

hence

$$f = \frac{1}{2\pi\sqrt{LC}}$$

$$= \frac{1}{2\pi\sqrt{\frac{100}{10^6} \times \frac{250}{10^{12}}}}$$

$$= \frac{1}{2\pi\frac{\sqrt{100} \times \sqrt{250}}{\sqrt{10^{18}}}}$$

$$= \frac{10^9}{2 \times \pi \times 10 \times 15.8}$$

$$= \frac{1000 \times 10^6}{992}$$

$$= 1 \times 10^6\text{Hz (approx)}$$

$$f = 1\text{MHz (approx)}$$

Example 23

What value of inductance is required in series with a capacitor of 500pF for the circuit to resonate at a frequency of 400kHz?

From the resonance formula

$$f = \frac{1}{2\pi\sqrt{LC}}$$

the inductance is

$$L = \frac{1}{4\pi^2 f^2 C}$$

Expressing the frequency and the capacitance in the basic units ($f = 400 \times 10^3$Hz and $C = 500 \times 10^{-12}$F)

$$= \frac{1}{4\pi^2 \times (400 \times 10^3)^2 \times (500 \times 10^{-12})}$$

Taking $\pi^2 = 10$

$$= \frac{10^{12}}{4 \times 10 \times 16 \times 10^4 \times 10^6 \times 500}$$

$$= \frac{1}{3200} \text{ H}$$

$$L = 310\mu\text{H (approx)}$$

Example 24

If the effective series inductance and capacitance of a vertical antenna are 20μH and 100pF respectively and the antenna is connected to a coil of 80μH inductance, what is the approximate resonant frequency?

The antenna and coil together resonate at a frequency determined by the capacitance and the sum of the antenna effective inductance and the loading coil inductance.

$$f = \frac{1}{2\pi\sqrt{LC}}$$

Here the relevant values of inductance and capacitance expressed in the basic units are

$$L = (20 + 80) \times 10^{-6}\text{H}$$

$$C = 100 \times 10^{-12}\text{F}$$

Therefore

$$f = \frac{1}{2\pi\sqrt{(100 \times 10^{-6}) \times (100 \times 10^{-12})}}$$

$$= \frac{1}{2\pi\sqrt{10^{-14}}} = \frac{10^7}{2\pi}$$

$$= 0.16 \times 10^7 = 1.6 \times 10^6$$

$$f = 1.6\text{MHz}$$

Example 25

An alternating voltage of 10V at a frequency of 5/πMHz is applied to a circuit of the following elements in series: (i) a capacitor of 100pF, (ii) a non-inductive resistor of 10Ω.

(a) What value of inductance in series is required to tune the circuit to resonance?

(b) At resonance, what is the current in the circuit?

(a) For the calculation of the inductance, the resistance can be ignored, since it has no effect on the resonant frequency, which is given by

$$f = \frac{1}{2\pi\sqrt{LC}}$$

rearranging

$$L = \frac{1}{4\pi^2 f^2 C}$$

Expressing the frequency and the capacitance in the proper units ($f = 5 \times 10^6/\pi$Hz and $C = 100 \times 10^{-12}$F)

$$L = \frac{1}{4\pi^2\left(\frac{25 \times 10^{12}}{\pi^2}\right) \times 100 \times 10^{-12}} \text{ henrys}$$

$$L = \frac{1}{10{,}000} \text{ H} = 100\mu\text{H}$$

(b) At resonance, the inductive and capacitive reactances cancel out and the circuit has a purely resistive impedance of 10Ω. The current I through the circuit at resonance can then be calculated directly from Ohm's Law: $I = V/R$. Since $V = 10$V and $R = 10\Omega$, the current at resonance is

$$I = \frac{10V}{10} = 1\text{A}$$

Example 26

A tuned circuit having a Q of 120 resonates at a frequency of 80kHz. What is its bandwidth at the –3dB level?

$$Q = \frac{f_r}{f_h - f_l}$$

$$120 = \frac{80}{f_h - f_l}$$

$$f_h - f_l = \frac{80}{120} = \frac{2}{3} \text{ kHz}$$

Bandwidth is approximately 666Hz.

Example 27

A coil of inductance 25μH has a resistance of 7Ω at a frequency of 2.5MHz. What is its Q at this frequency?

$$Q = \frac{2\pi f L}{R}$$

$$= \frac{2\pi \times 2.5 \times 10^6 \times 25 \times 10^{-6}}{7}$$

$$= \frac{125\pi}{7}$$

$$= 56$$

CHAPTER 3

Solid-state devices and valves

The majority of electronic circuits found in amateur equipment today utilise solid-state devices. These devices are based on 'semiconductor' materials; these can be broadly defined as materials whose resistance lies between that of a conductor and an insulator. The most common materials in this category, silicon and germanium, are the basis of the majority of transistors and diodes.

The silicon (Si) atom

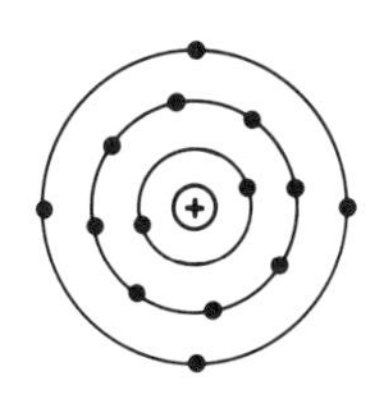

Fig 3.1. The silicon atom

A simplified representation of a silicon atom is shown in Fig 3.1. Around the positive nucleus there are three rings (orbits or shells) containing negatively charged electrons. The sum of the negative charges balances the positive charge of the nucleus – thus the atom is electrically neutral. The first ring contains two electrons and cannot accept any more. The next ring contains eight electrons. However, the third ring contains only four electrons and these join with the four electrons in the outer rings of adjacent atoms to form a crystal lattice. The outer electrons are not very far away from the nucleus and are not free to move from the lattice. Thus pure silicon is a good insulator.

The germanium (Ge) atom

The germanium atom has a total of 32 electrons in four rings which from the centre outwards contain two, eight, 18 and four electrons. The four electrons in the outer ring, as in the case of the silicon atom, join with those of adjacent atoms to form a crystal lattice. Because the outer ring is one further away from the nucleus, they can become detached more easily than in the silicon crystal.

Semiconductor materials

The manufacturing process of solid-state devices is complex: it requires the refinement of silicon and germanium to an extremely high purity and then the introduction of a very small but closely controlled amount of an impurity. It is this impurity that gives the base material (silicon or germanium) its semiconductor properties which provide transistor and diode action.

N-type material

In the manufacturing process, impurity atoms having an outer ring containing five electrons are introduced into the crystal lattice. This process is known as 'doping' and the resulting material is known as 'N-type'. Typical impurities are phosphorus and arsenic.

P-type material

If the impurity added has only three electrons in its outer ring then a gap is left in the lattice which could be filled by a free electron. Such a gap is called a 'hole' and the resulting material is known as 'P-type'. Typical impurities are boron and aluminium.

The doped material of both N- and P-types is electrically neutral because each one of the individual atoms present is itself electrically neutral.

The PN junction

A diode consists of a small single piece of silicon or germanium in which one end has been made N-type and the other P-type. Because the two ends of the material have different characteristics as described above, there is a diffusion of holes in one direction and electrons in the other. This forms an area where electrons will have jumped into vacant holes and a boundary or junction formed stopping any further migration of holes and electrons as shown diagrammatically in Fig 3.2. This area is known as the 'depletion layer'. This layer is typically 0.001mm in thickness.

Fig 3.2. The junction diode

The depletion layer has formed a region where work must be done to get further electron/hole movement, ie energy must be provided externally. This can be provided in the form of a voltage from, say, a battery.

Consider the circuit of Fig 3.3 with the polarity of the battery as shown. The electrons from the N-region will be attracted by the positive plate of the battery and so will be assisted to cross the depletion layer. This will then leave holes in the N-region which are filled by electrons entering from the wire conductor and the rest of the circuit by a chain reaction. The electrons that enter the P-region will be pulled into the wire due to this chain reaction. Thus a conventional current flows (note that conventional current flow is in a direction opposite to that of electron flow). This condition is known as 'forward bias'.

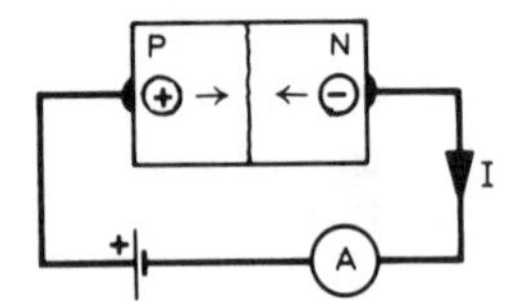

Fig 3.3. The forward-biased junction diode

If the polarity of the battery is reversed it is found that the electrons accumulate at the battery end of the N-region and similarly for the holes in the P-region.

Thus negligible current flows. This is known as 'reverse bias'. The device that has been formed is known as a 'diode' (or 'rectifier'). It allows current to flow in only one direction and this is determined by an externally applied voltage. In the forward bias condition it is found that no current flows until a certain voltage has been applied across the diode (called the 'barrier potential'). With germanium this is found to be about 0.3V and with silicon about 0.6V. This corresponds to the work that must be done in helping the electron to cross the depletion layer. The depletion layer becomes narrower with increased forward bias.

This property of the PN junction can be described by the graphs in Fig 3.4. A small current is found to flow under reverse bias conditions and increases with temperature. This is much more marked in germanium diodes than silicon diodes. This current is known as 'leakage current'. Please note the change in scale of the vertical axes.

This basic diode is used for the rectification of alternating voltages in power supplies, demodulating signals in radio receivers and certain logic functions. The type of diode for an application will depend on current, maximum reverse voltage to be expected and frequency.

By changes in the manufacturing process, diodes can be made to exhibit somewhat different characteristics. Two of these – the zener diode and the variable capacitance diode – are examined next.

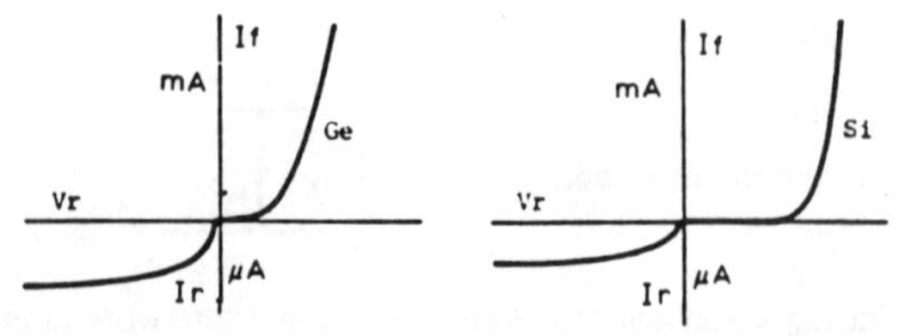

Fig 3.4. Characteristics of germanium and silicon junction diodes

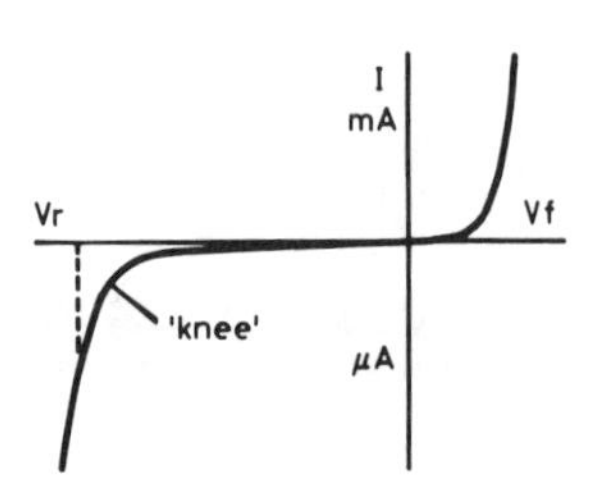

Fig 3.5. Characteristic of the zener diode

Zener diode

In normal diodes only a small current will flow under reverse bias conditions until the voltage becomes so large that there is a catastrophic breakdown of the diode and it is of no further use. With the zener diode the manufacture is such that only a small current flows up to a certain well-defined point, when there is a sudden increase in current and the voltage across the diode becomes virtually constant. Typical characteristics are shown in Fig 3.5.

There are technically two diodes, the zener and avalanche, depending on voltage, but by popular usage 'zener' is used to describe both types. These diodes are manufactured in various fixed values from less than 3V up to about 150V. They can be put in series for greater values. Zener diodes are useful in DC power supplies which need to provide a constant output voltage irrespective of loading and changes in input voltages.

Variable-capacitance diode

This diode is also referred to as a 'varactor diode' or more commonly as a 'varicap diode'. It is used in the reverse bias mode and uses the width of the depletion layer to vary the capacitance. All diodes exhibit this effect to some extent but the doping of the semiconductor used can have a marked effect on the capacitance range available. The depletion layer can be likened to a parallel-plate capacitor – by widening the distance between the plates the capacitance drops and vice versa. As the reverse bias increases so the depletion layer widens and the capacitance decreases. This is depicted in Fig 3.6 for a BB110-type diode.

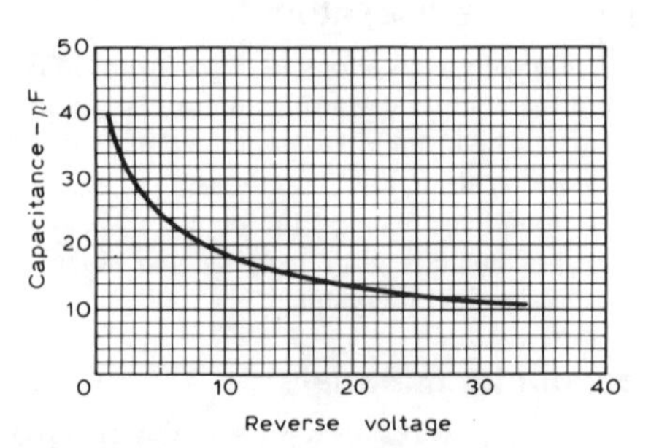

Fig 3.6. Variation in capacitance of a varactor diode

There are many different devices available with varying specifications and in total cover the range from 2pF up to 600pF. Their use varies from tuning circuits to FM modulators and frequency multipliers.

NPN and PNP transistors

The word 'transistor' must be treated carefully. It is in fact a word describing a family of devices but is often used to refer to the NPN and PNP type only. The NPN and PNP transistors are technically 'bipolar transistors'.

The transistor was discovered in 1947 and entered production in 1951. Early types were made from germanium but silicon has now surpassed them in most applications. It has also been fabricated in various forms, and names such as 'junction', 'epitaxial', 'planar' etc are common in data sheets. The basic operation is, however, the same, the different methods of fabrication offering different characteristics.

The device is made from a single piece of silicon (or germanium). The various regions are created chemically to give an NPN or PNP structure – that is two PN junctions (the NPN form being more common). It has three terminals known as 'collector', 'base' and 'emitter'. Fig 3.7 shows a typical NPN junction-type bipolar transistor. The device is 'current controlled', ie the base current controls the amount of current flowing between the collector and emitter. The higher the base current, the greater the current flow between collector and emitter and of course the converse.

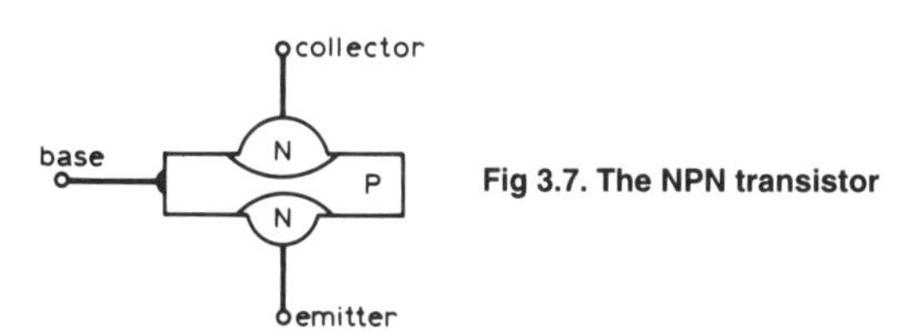

Fig 3.7. The NPN transistor

To enable a bipolar transistor to function, the voltages of the collector, base and emitter must be correct with respect to one another – this is known as 'biasing'. The following discussion is for a silicon device as these are more common. Also, the description is for an NPN device – for a PNP type all voltages and currents are reversed. Fig 3.8 shows a very basic method of biasing; in reality only one power supply source would be used.

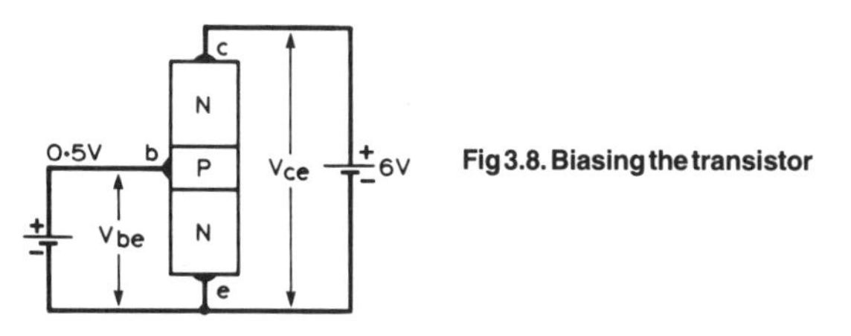

Fig 3.8. Biasing the transistor

From Fig 3.8 it can be seen that the base is 0.5V above the emitter, and this therefore forms a forward-biased PN junction. The collector is higher than the base voltage and so this

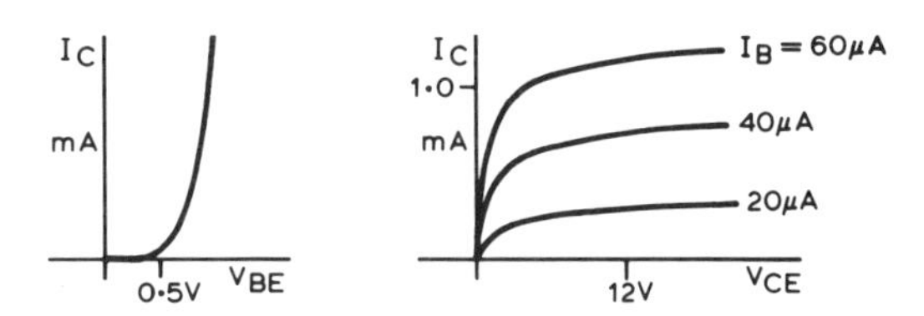

Fig 3.9. Characteristics of a small-signal transistor

forms a reverse-biased PN junction. A set of measurements could be carried out for various values of collector voltage, collector current, base voltage and base current and a set of curves would be obtained as shown in Fig 3.9. These are known as the 'characteristic curves'. They are the graphs shown in transistor data sheets.

There are various parameters available to describe a transistor. These can relate to voltage, current, power, gain, frequency and operating temperature; which of these are relevant depends on the application. The voltage rating gives an indication of the maximum voltages the transistor can withstand; the current will specify maximum currents the device can handle; and power, the maximum power the device can deliver to a load or dissipate itself. The gain is a relationship between input and expected output, the frequency will specify the range the device is capable of working over and the temperature how hot the device can run (this is also related to power). Transistors come in various shapes and sizes, usually depending on the current or power to be handled. Unfortunately there are subdivisions of the above which can make matters more complicated but some basic parameters are discussed below. The maximum collector voltage specifies, as the name suggests, the maximum voltage that the collector can withstand, hence the DC power supply may help determine the choice of transistor.

The current gain relates the change of collector current with change of base current. If the collector current changes from 1 to 2mA when the base current changes from 5 to 10μA, then the current gain is 1mA/5μA = 200. The symbol for current gain is h_{FE} or, mainly in older literature, β. Depending on transistor type, current gains from 2 to 800 are possible. Values for silicon devices are higher than those for germanium.

The transition frequency f_T is the frequency at which the current gain of the device has fallen to unity. Transistors are normally used up to a value of 10–15% of this figure. There are no hard and fast rules as this depends on the application.

The power the device will handle is dependent on several factors: how efficient the particular circuit arrangement is, the output power wanted, the ambient temperature and the cooling available (heat sink, air circulation etc). Always run as cool as possible for reliability. A typical example of the advantage of a heat sink can be shown by considering a BFY50. This will dissipate 0.8W with no heat sink if the temperature can be kept below 25°C. With a given size heat sink, this could be extended to 100°C and a dissipation of 2.8W. It is the actual temperature of the chip, ie the small piece of silicon on which the transistor is assembled, that is of paramount importance.

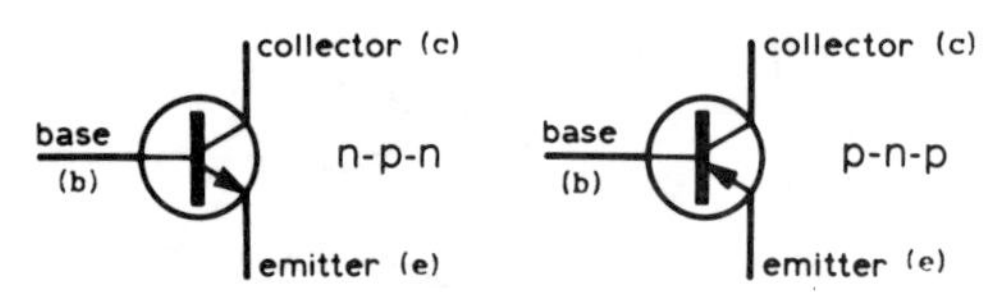

Fig 3.10. Transistor symbols

Germanium transistors were developed before the silicon transistor but they tend to be much more affected by operating temperatures. For example, the maximum safe working temperature for a germanium device is about 75°C while for a silicon device it is about 150°C. Hence silicon devices are generally preferred.

The British Standards (BS) symbols for the two types of transistor are given in Fig 3.10. The arrow is always on the emitter.

Field-effect transistor (FET)

This is another member of the transistor family which is widely used and has advantages in certain applications. It comes in various forms: JFET, IGFET, MOSFET etc, but only the JFET (junction FET) will be described here. Both N- and P-types exist but the description of operation will be that of an N-channel type. Fig 3.11 shows the construction diagrammatically.

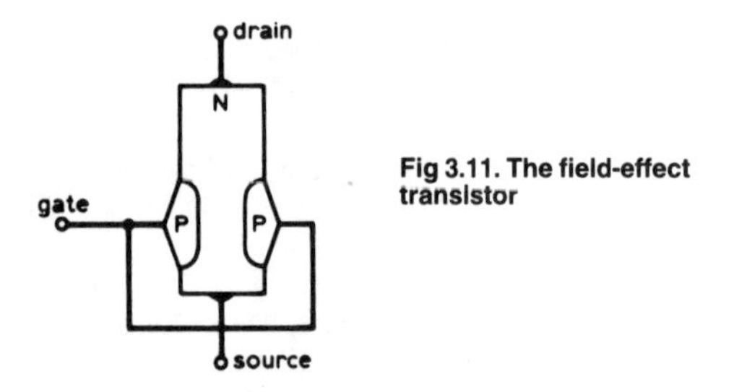

Fig 3.11. The field-effect transistor

It consists essentially of a P-type region diffused into an N-type channel. The P-type region, or 'gate', creates a junction with the N-type bar and hence a depletion layer. The main current flow is from drain to source, the voltage applied at the gate being used to control this current. Thus, the FET is a voltage-controlled device as opposed to the NPN/PNP transistor which is current controlled (ie base current controls collector-emitter current).

To operate, biasing is required (see Fig 3.12). This is

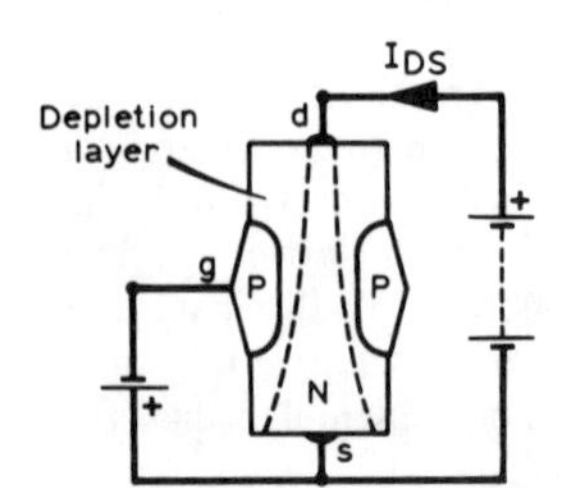

Fig 3.12. Configuration and operation of a FET

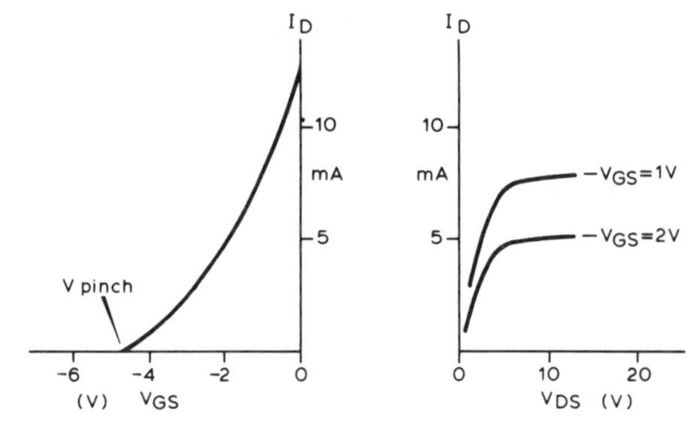

Fig 3.13. Characteristics of a 2N3823 FET

shown with batteries but in reality only one DC supply would be used. This must make the gate negative with respect to the source and hence a reverse-biased PN junction, thus exhibiting high input resistance. The drain must be positive with respect to the source. By varying the voltage on the gate the depletion layer width can be varied – the more negative the voltage, the wider the depletion layer, and the less the drain-source current. The voltage required to reduce the drain current to zero is known as the 'pinch-off voltage'. A typical set of characteristics for a 2N3823 JFET is shown in Fig 3.13.

The FET is a useful device as it always exhibits very high input resistance. The BS symbols for the types of JFET are given in Fig 3.14.

Application of solid-state devices

The application of solid-state devices is almost endless. Individual devices such as the transistor and diode are called 'discrete semiconductors'. The following gives a sample of the uses and typical circuit arrangements. These are sufficient for the examination. However, there are many other arrangements and these can be found in many of the amateur radio handbooks available. All semiconductor devices can easily be damaged by excessive power, voltage and/or current. Power devices should be mounted on heat sinks – that includes diodes as well as transistors. Never impede the cooling process. That said, and if used within their design limits, semiconductors are very reliable.

The diode or rectifier

It is hard to define the difference between these two terms – usually 'diode' is used for low-power applications and 'rectifier' for higher-power mains applications. The silicon diode

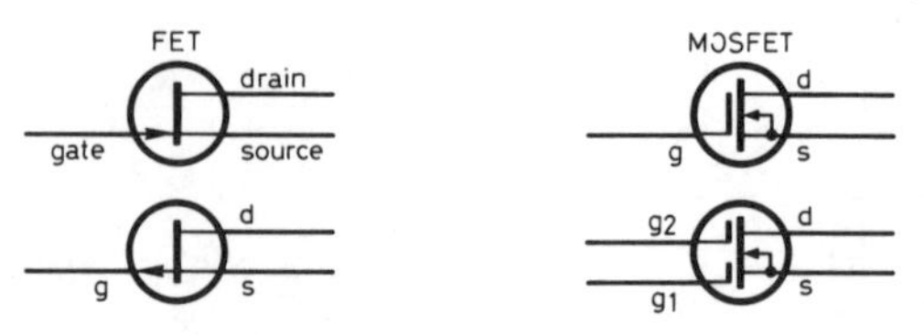

Fig 3.14. Symbols for various FETs

has now replaced the germanium diode for mains rectification. The germanium diode may still be found in signal-detection circuits where its low forward voltage drop is advantageous. The main use of the rectifier in power supples is dealt with in Chapter 6.

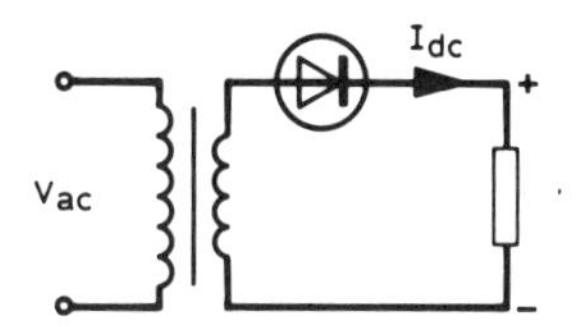

Fig 3.15. Use of the diode as a rectifier

Fig 3.15 shows a typical use as a rectifier. It will produce half-wave rectification as indicated. The circuit of Fig 3.16 shows the typical use of the diode as an AM demodulator, the tuned circuit being the output from the final IF stage. A similar arrangement can be used to generate AGC voltages. The diode can also be used in FM demodulation (see Chapter 5).

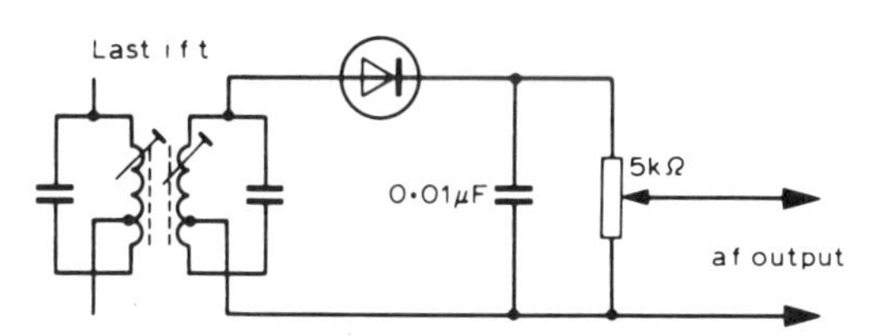

Fig 3.16. Use of the diode as a demodulator

Zener diode

The basic arrangement is shown in Fig 3.17. The resistor R_S is chosen so that when the load resistor R_L is carrying its normal current, the voltage drop across R will ensure that the diode is always operating beyond the knee of its characteristic (see Fig 3.5). If the output voltage tends to rise due to a reduction in load current, the zener diode will take more current from the supply and so keep the output voltage reasonably constant. The zener diode can also be used as a basis for more sophisticated power supplies (see Chapter 6).

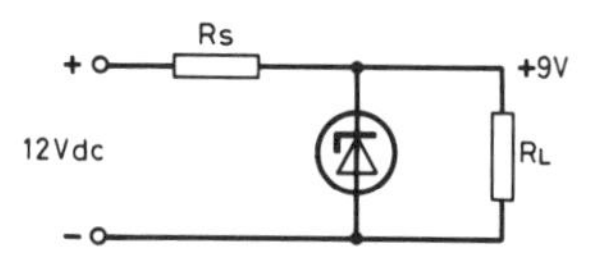

Fig 3.17. Simple zener diode regulator circuit

Variable-capacitance diode

This is used where the frequency of a tuned circuit must be varied. Its use is mainly at RF frequencies. The nominal operating point of a varactor diode is determined by its standing bias voltage, and this must keep it in reverse bias and near the midpoint of its characteristic so that equal changes in applied voltage cause equal changes in capacitance. Fig 3.18 shows a typical application.

The resonant frequency of the tuned circuit is determined by C in parallel with C1 and C_V in series. A standing bias is applied via R1 and the capacitance of the whole circuit, and

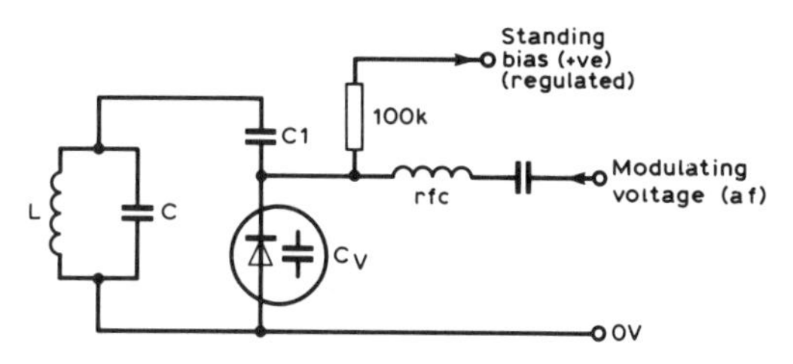

Fig 3.18. Typical variable-capacitance diode application

hence resonant frequency, is altered by the voltage applied via RFC. If an audio voltage is applied via RFC then the resonant frequency will change in sympathy with this and, if the tuned circuit is part of an oscillator, frequency modulation will be generated. The application of a DC voltage from a potentiometer instead of an audio input provides variable tuning. Further discussion will be found in Chapter 4. The varactor diode in a similar arrangement forms the heart of the frequency synthesiser. The varactor diode can also be used for frequency multiplication, eg 144MHz to 432MHz.

The transistor as a small-signal amplifier

Small-signal amplifiers are so called because their inputs are usually in the microvolt or millivolt range as opposed to large signal amplifiers which have inputs of the order of volts. The latter are normally referred to as 'power' amplifiers; these are dealt with later. Whichever type, the base, emitter and collector of the transistor must be set to various DC voltages so that it will act as an amplifier. This setting up of DC conditions is known as 'biasing' and if incorrectly chosen the amplifier may distort the signal. Amplifiers can be used for DC, through AF, up to RF and microwaves. Only AF and RF types are considered.

Common-emitter amplifer

Fig 3.19 shows a typical common-emitter circuit and represents the most widely used of the amplifiers. Capacitors C1 and C2 provide AC coupling into and out of the circuit respectively. R1 and R2 provide biasing (note that this is not the only method of biasing but is the most stable). R3 is the collector load while R4 helps set up the DC bias conditions with C3 (the emitter bypass capacitor) acting as an AC short-circuit across it. With C3 acting as an AC short-circuit (ie its

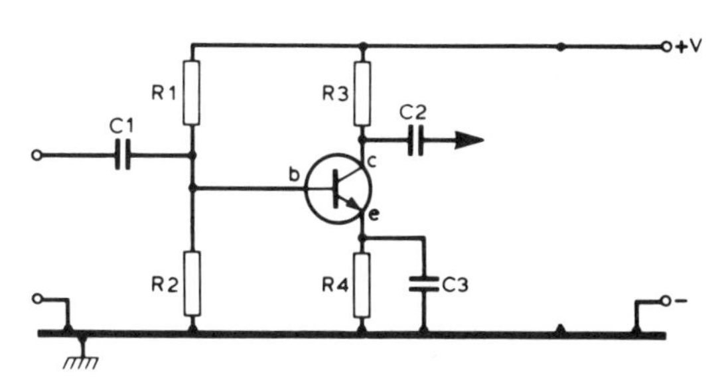

Fig 3.19. Typical common-emitter amplifier

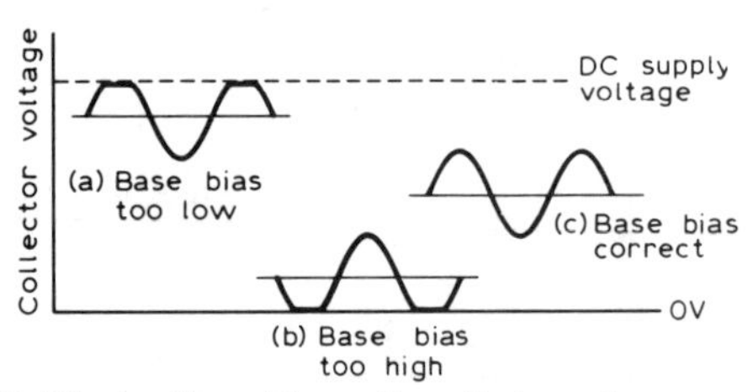

Fig 3.20. Effects of base bias on the collector voltage

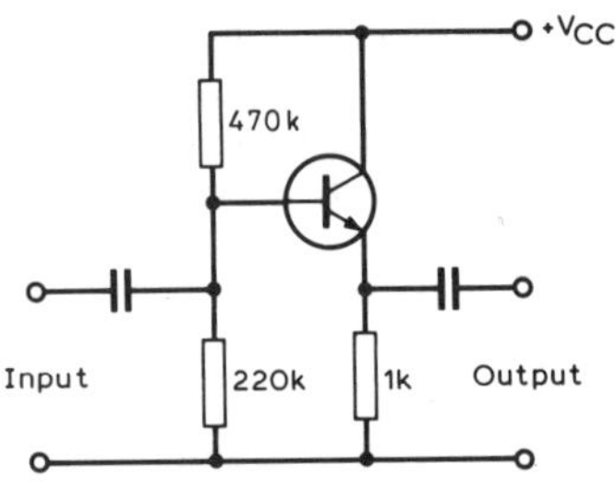

Fig 3.21. Typical emitter-follower circuit

reactance is very small at the operating frequency), the emitter, as far as the AC is concerned, appears the same as the 0V line and is thus common to input and output signals. This gives the amplifier its name.

The values of C1, C2 and C3 are chosen so that they have very low reactance at the operating frequency. Their value will therefore vary according to the frequency being amplified; at audio they may be electrolytic capacitors, while at RF they may be in the 1000pF range or less. The biasing of the transistor has to be carefully chosen so that minimum distortion will occur of the signal being amplified. The voltage of the base must be about 0.6V above the emitter (assuming a silicon transistor). This allows a standing bias current through the transistor determined by R4. Knowing this value means the standing (or quiescent) voltage of the collector is known. Fig 3.20 shows the effect of too high and too low a voltage at the collector. In cases (a) and (b) distortion occurs which manifests itself as harmonics. Case (a) is when the input voltage cannot maintain the 0.6V differential between the base and emitter and has effectively caused zero collector current; the transistor is 'switched off'.

The operation of a common-emitter circuit is as follows. Assuming the input voltage at the base is converted to a base current, then if the input voltage increases, so does the base current. This causes an increase in collector current and because one end of R3 is attached to +V, then for increased current the voltage across R3 must increase and so the voltage at the collector decreases. If the input voltage decreases, then so does the base current and hence the collector current. This causes less current through R3 and so the collector voltage rises. It should be noted that the collector voltage does the opposite to the input voltage, ie it is out of phase or inverted (ie 180°). Typical values for R1, R2, R3 and R4 might be 27kΩ, 8.2kΩ, 4.7kΩ and 1.5kΩ respectively.

One aspect not yet considered is how one circuit is affected by another one connected to it. For example, if one common-emitter amplifier cannot give enough gain then it can be followed by a second one in series. All electronics are the effective connection of a series of basic building blocks. The second circuit obviously must take a signal from the first one and this leads to the question of how one circuit 'loads' another. This is commonly expressed by the concept of input and output impedances.

Ideally the input resistance should be high so that it does not load a previous circuit too much and the output resistance should be reasonably low so that most of the output voltage can be passed to the next circuit. For a typical common-emitter amplifier the input resistance is between 500Ω and 2kΩ and the output resistance between 5 and 20kΩ but the circuit resistors do have some effect on these figures. The common-emitter amplifier does not always represent an ideal arrangement and there are two other transistor configurations to consider – the common-collector (or emitter-follower) and common-base circuits.

Common-collector (emitter-follower circuit)

The name in brackets for this is the most popular, but the first one follows a logical sequence. It may also be referred to as a 'buffer'. Fig 3.21 shows a typical circuit with values.

If the input voltage increases in this circuit, the base current rises and hence the collector current. As the emitter current is almost the same as the collector current, then the current through the emitter rises and hence the voltage at the emitter rises (one end of R3 is fixed to 0V and cannot change). If the input voltage decreases then the emitter voltage must also decrease with like reasoning. Because the base is always 0.6V greater than the emitter, then the emitter voltage is always less than the base voltage; ie there is a voltage gain of less than unity. The input resistance of this circuit is about 100kΩ while the output resistance is maybe only 1kΩ (again, both are dependent on the effects of circuit resistors). This makes an ideal circuit in preventing loading of one circuit and feeding the next one with little voltage loss. It acts as a buffer between two circuits and hence its alternative description.

Common-base circuit

This is the third configuration and Fig 3.22 shows a typical arrangement. The signal is applied between base and emitter. If the input voltage increases, then the emitter voltage will rise

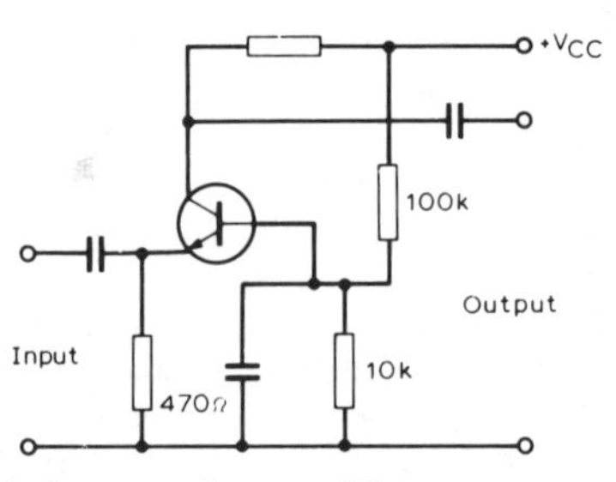

Fig 3.22. Typical common-base amplifier

Table 3.1. Characteristics of transistor configurations

	Common emitter	Common collector	Common base
R in	1kΩ	100kΩ	50–100Ω
R out	20kΩ	1kΩ	50kΩ
Current gain	50	50	0.98
Voltage gain	−100	0.98	100

and cause less of a differential between emitter and base. This means that the transistor takes less base current and so the collector current drops. This in turn causes a rise in collector voltage. If the input voltage drops then the reverse happens and the collector voltage drops. Hence, the collector voltage is in phase with the input voltage and so there is no phase inversion for voltage. The input impedance is very low, about 50–100Ω, so it is easy to match to a transmission line and/or antenna system. The output resistance is very high (at least 50kΩ).

Each of these circuits just discussed has its own specific properties as mentioned, and these are summarised in Table 3.1. It is stressed that these values are only typical as so much depends on other factors such as external resistors, transistor parameters and operating conditions. The negative sign in front of a gain figure shows that the output signal is out of phase with the input signal.

As the frequencies in use rise, so the internal capacitance effects of the transistor must be taken into account and the word 'resistance' in the above discussion must be replaced by 'impedance'. Generally the input impedance gets lower and so does the output impedance. The value of the internal capacitance eventually has to be taken into account when designing adjacent circuitry but this is a matter above the present level of discussion.

Power amplifiers

This is the case of the large-signal amplifier. Power amplifiers are needed for both AF and RF applications. They are required to provide power to a load such as a loudspeaker or antenna. The basic circuits will be similar and the operation is the same as the small-signal amplifier but there will be detail differences. There are also variations in order to minimise power dissipation.

Class A operation

The common-emitter amplifier considered earlier provides an exact, but amplified, replica of the input signal – both positive and negative half-cycles of the input are amplified equally. This is defined as 'Class A' operation and can be shown theoretically to have a maximum efficiency of 50%. At low power levels this is not important but if 50W is wanted in the load, then the transistor must dissipate 50W and the power supply has to provide 100W. This would be the ideal case; in practice the amplifier is more likely to be 40–45% efficient. This then represents a high power wastage and of course means the provision of adequate heat sinks. The advantage of Class A operation is low distortion (ie low harmonic content). A Class A power amplifier will look very similar to Fig 3.19. The circuit in Fig 3.23 shows the case when the collector resistor is replaced by a transformer feeding a loudspeaker. This arrangement allows optimum power matching.

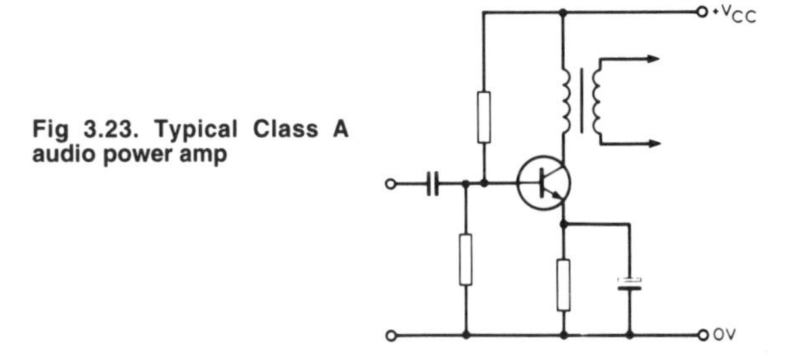

Fig 3.23. Typical Class A audio power amp

Efficiency and power wastage are a serious disadvantage of Class A and ways have been found to overcome this. The application will determine the approach. There are two other classes of operation – Class B and Class C.

Class B operation

This amplifier only amplifies half a cycle but theoretically is 78% efficient. In practice it should give at least 66% efficiency. Fig 3.24 shows a typical circuit which as can be seen looks identical to a Class A amplifier. The difference is that the bias resistors are chosen so that the base is only just at about 0.6V; as stated earlier, if it is taken lower than this the transistor stops passing current. On the positive half-cycle the base current will increase, as does the collector current, and an inverted (but amplified) half-sinewave is produced. On the negative half of the input signal, the base is taken to less than 0.6V and so the transistor will not pass current (switches off) and does not amplify. Thus only half of the input signal is amplified. For audio applications a Class B push-pull amplifier is used, as shown in Fig 3.25, while at radio frequencies the collector load is replaced by a tuned circuit.

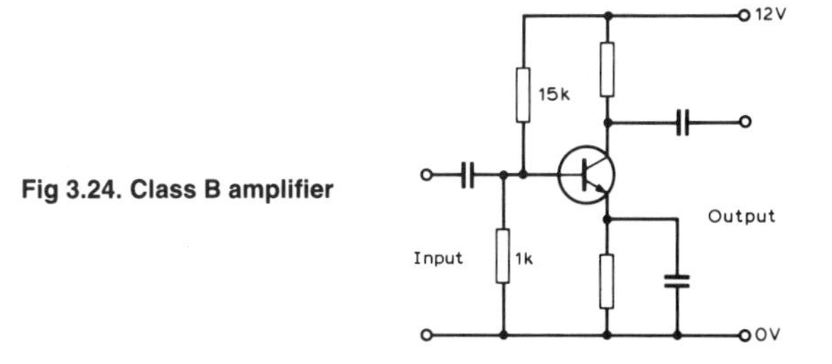

Fig 3.24. Class B amplifier

In Fig 3.25 the top transistor amplifies the positive half-cycle and the other the negative half-cycle. They are added together at the output. This is depicted in Fig 3.26. One problem that can occur is that the two halves of the sinewave do not match exactly due to differences in the transistors, and

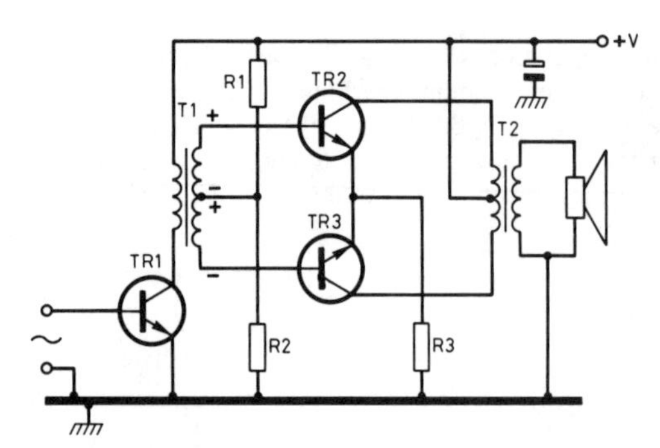

Fig 3.25. Class B push-pull circuit amplifier

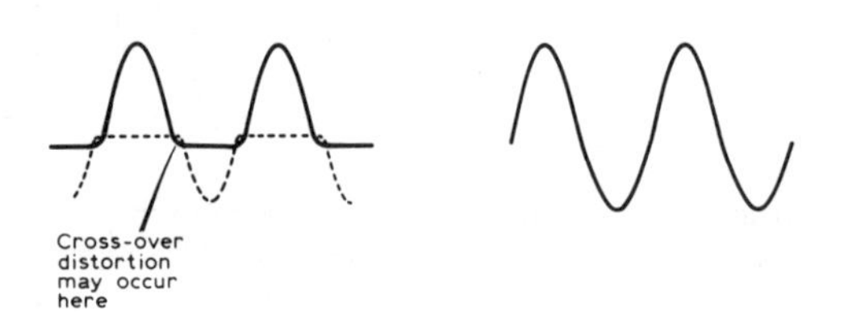

Fig 3.26. Signal combination in Class B amplifier

crossover distortion occurs. To minimise this, some forward bias is applied so that the transistor is passing a small current with no input – this is Class AB operation.

Class C amplifiers

This amplifier amplifies less than half a cycle. A typical circuit is given in Fig 3.27. It will be noticed that there is no bias resistor from base to positive line, only that from base to ground.

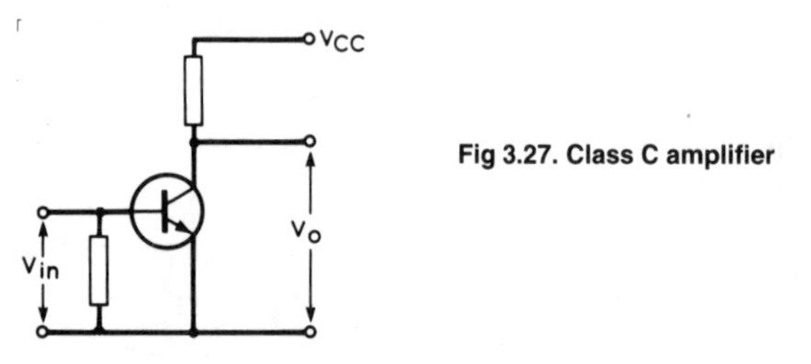

Fig 3.27. Class C amplifier

The transistor is kept non-conducting by this latter resistor. For the transistor to conduct and hence amplify, the base must be taken from 0 to 0.6V initially (ie part of the half-cycle has already started); the collector current then increases rapidly as the base becomes more positive. As the input decreases the transistor will switch off before the half cycle reaches 0V (ie at 0.6V). On the negative half-cycle of input, the transistor remains firmly in a non-conducting state. The efficiency of this amplifier cannot be quoted exactly as it depends on how much of the half-cycle it conducts for. However, it will be be more efficient than a Class A or possibly Class B amplifier – assume about 66%. In a typical Class C transmitter the collector load is replaced by a tuned circuit. As for the Class B stage, the energy stored in the tuned circuit supplies the other half-cycle.

The transistor as a switch

The Class C amplifier leads straight into the use of the transistor as a switch. In this case the criterion becomes 'is the transistor conducting or not?'. This is the basis of digital circuits, ie ON or OFF; no intermediate state is wanted. The Class C amplifier is also representative of a transistor switch circuit, as shown in Fig 3.28.

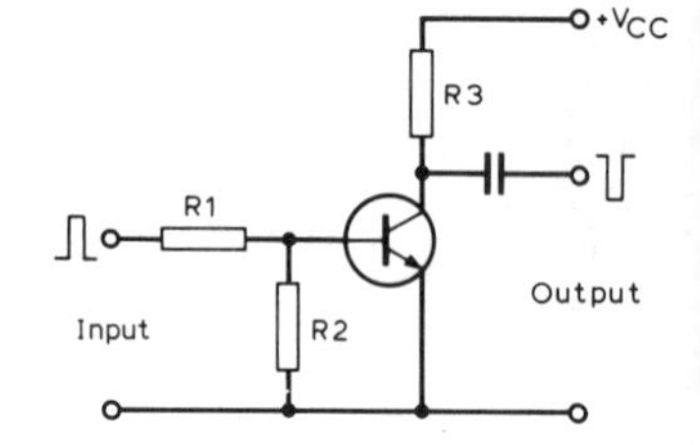

Fig 3.28. Typical transistor switch circuit

The operation of the circuit is similar to that of the Class C amplifier, however, the input is usually a rectangular pulse and not a sinewave. The resistor R2 keeps the transistor in a non-conducting state and R1 purely limits the base current. If the input signal exceeds 0.6V (assuming a silicon transistor) the transistor suddenly conducts and the output voltage, because of R3, drops to a low value, ie the signal is inverted. When the input falls to less than 0.6V then the transistor will no longer conduct. The collector resistor R3 could be replaced by a lamp or relay so that when an input signal is present current flows and the relay operates or the lamp lights. With no input signal the transistor will not conduct.

The transistor as a tuned amplifier

If the collector resistor of an amplifier is replaced by a parallel-tuned circuit, an amplifier can be created that will only amplify a selected range of frequencies. A typical circuit is shown in Fig 3.29.

From earlier theory in Chapter 2, we know that the dynamic resistance of the parallel-tuned circuit is a maximum at resonance. Maximum output of this circuit will occur at this

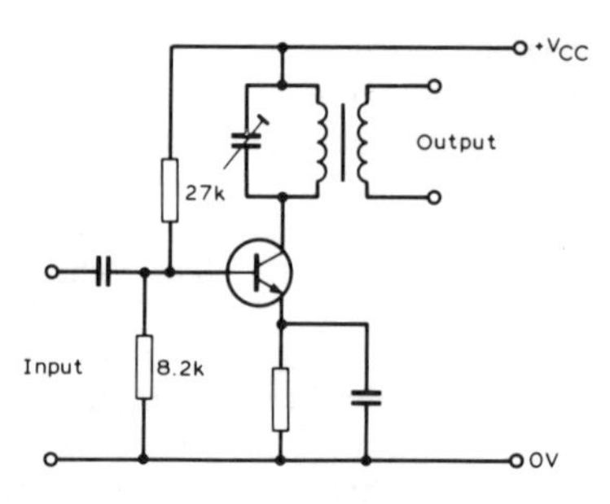

Fig 3.29. Typical tuned amplifier circuit

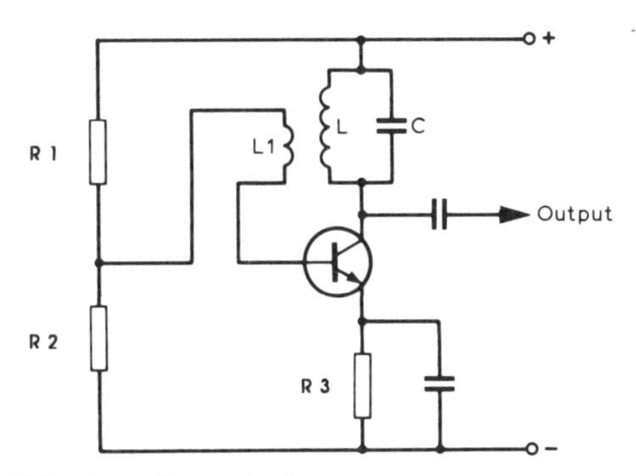

Fig 3.30. Basic oscillator circuit

point. Going either side will give reduced output, ie maximum amplification occurs at the resonant frequency of the tuned circuit.

If the parallel resistance across a tuned circuit decreases, then this lowers the dynamic resistance and so degrades the performance of the tuned circuit, widening the band of frequencies amplified. The input impedance of a following circuit appears across the tuned circuit. To minimise this degrading of the amplifier response, the output is usually taken from a secondary winding close to the tuned inductor in the collector circuit, ie it becomes a transformer with an untuned secondary winding (see Chapter 2).

If in Fig 3.29 the bias on the transistor is changed to Class C, the collector current is distorted and contains second and third harmonics. Thus, if the tuned circuit is resonated at one of these, the circuit would operate as a frequency doubler or tripler.

The transistor as an oscillator

At some point in a receiver or transmitter it is necessary to produce a continuous sinewave. This is accomplished by a circuit known as an 'oscillator' and Fig 3.30 shows a basic circuit. The tuned circuit L-C determines the frequency of oscillation; the two coils L and L1 form the primary and secondary windings of a transformer; resistors R1, R2 and R3 set the bias conditions. The circuit is similar to a common-emitter amplifier that produces a 180° phase shift between input and output voltages.

When the circuit is switched ON there is a momentary surge which causes the circuit L-C to try to oscillate. This signal is fed back by the transformer action of L and L1. The phasing is such that this will cause a signal at the collector which adds to the original and so increases it. This is fed back again in the correct phase and so oscillations are maintained. Thus the principle of the oscillator has been established.

Oscillators are amplifiers in which there is intentional positive feedback from the output to the input. This feedback can be achieved in several ways, and hence there are various oscillators, usually named after their originator, eg Colpitts, Clapp-Gouriet, Hartley, Vackar, etc. The first two are the most common in amateur radio. In general there is little to choose between them.

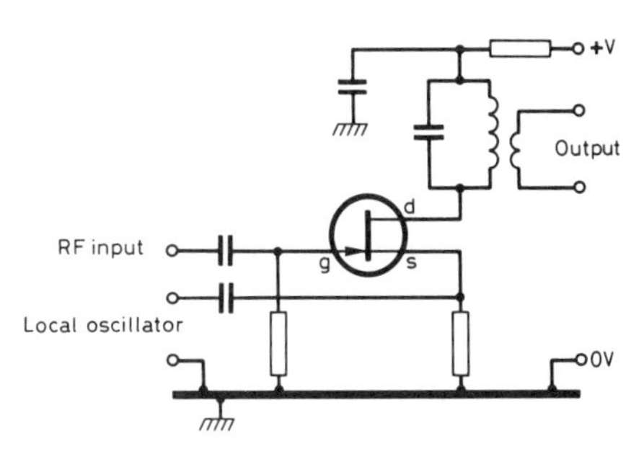

Fig 3.31. Typical junction FET mixer circuit

When one end of the tuned circuit which determines the oscillator frequency is earthed, it can be replaced by a quartz crystal, hence the 'crystal oscillator'. The oscillator frequency is now determined by the crystal and the frequency stability is therefore very much higher, but the oscillator cannot be tuned (see Chapter 2).

The transistor as a mixer

Mixer circuits are used to change from one frequency to another and these rely on the non-linear property of a circuit. Mixers produce a range of output frequencies and the required output is chosen by means of a tuned circuit or filter. The circuit in Fig 3.31 is the typical application of a junction FET as a mixer.

The local oscillator is fed in on the source while the RF signal is fed in, often by a tuned circuit, to the gate. These signals mix and the wanted mixing product is picked out by the tuned circuit in the drain circuit. The advantage of using a transistor or FET as a mixer is that it can provide gain at the same time.

The dual-gate MOSFET is also now very common as a mixer and a typical circuit is given in Chapter 5 (Fig 5.4).

Integrated circuits

These arise from the extension of transistor fabrication techniques to the assembly of complex electronic circuits on a single silicon chip. Thus most low-power stages used in receivers and transmitters, such as amplifiers (RF, IF and AF), demodulators, oscillators, stabilisers etc, are available. Few peripheral discrete components are required. Thus equipment has become smaller and more complex with extra facilities, and at the same time more reliable.

Introduction to valves

Valves function by the thermionic emission of electrons (negative) from a heated cathode in a vacuum. The electrons are attracted to an anode which is maintained at a positive potential. Thus electrons flow from the cathode to the anode, but the 'conventional' current is considered to flow in the opposite direction.

The cathode may be 'directly' heated, in which case it is a tungsten wire which is heated by the passing of a current

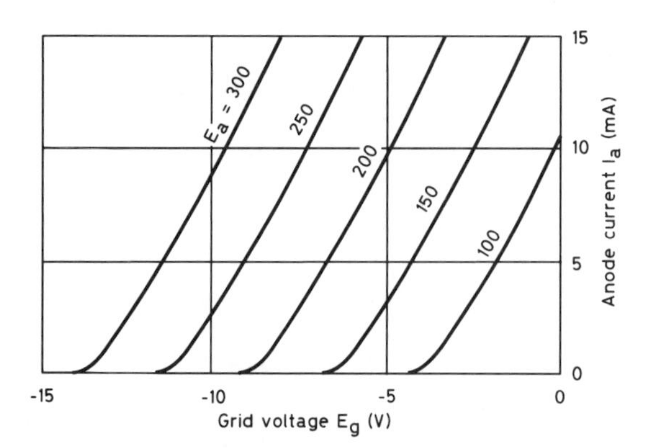

Fig 3.32. Anode current/grid-voltage characteristics

through it. An 'indirectly' heated cathode consists of a thin tube which is coated with various compounds of barium or thorium and is maintained at the required temperature by a heater which is inside and insulated from the cathode tube.

The basic two-electrode valve or 'diode' is a one-way device, ie it can rectify an alternating current. This application has now been taken over by the solid-state diode.

The addition between the cathode and the anode of a metal grid (control grid g_1), maintained at a negative potential, enables the electron current to be changed by varying the voltage applied to the latter. Thus the anode current is controlled. The relationship between anode current (I_a) and grid voltage (E_g) (negative) is shown in Fig 3.32, while Fig 3.33 shows how a valve functions as an amplifier. The anode load is a tuned circuit in an RF amplifier. The similarity to the transistor situation should be noted; it must be emphasised that a valve is a high-impedance device, whereas the transistor is of low impedance.

The terms 'Class A' etc are used to indicate the proportion of the input waveform over which the anode current flows. In Class A anode current flows over the whole of the input wave, and so on. Again the similarity to the transistor situation should be noted.

A second grid, the screen grid (g_2) can be located between the control grid and the anode, and such a valve is termed a 'tetrode'. The anode current is now much less dependent on the anode voltage: see Fig 3.34. The amplification and the impedance are increased, but the capacitance between the control grid and the anode is reduced from the order of 5–10pF to the order of 0.02pF. This is particularly significant because in a triode the higher grid/anode capacitance is in

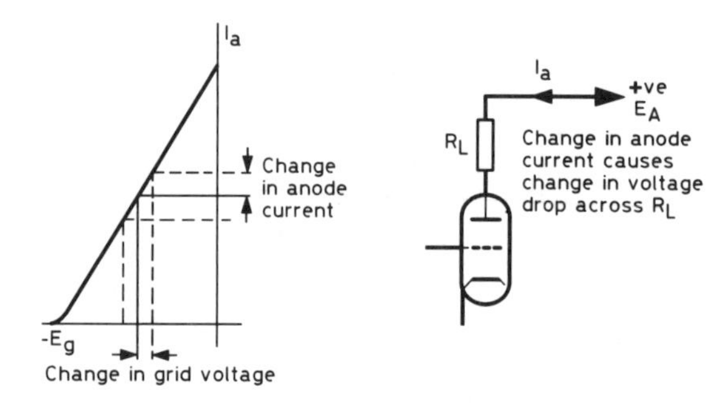

Fig 3.33. How the triode functions as an amplifier

effect connected between the output and the input of the amplifier circuit. Thus there is sufficient positive feedback to cause oscillation. This capacitance must be 'neutralised' by a small external capacitor. The low capacitance of the tetrode is unlikely to cause oscillation unless the mechanical layout of the circuit is bad.

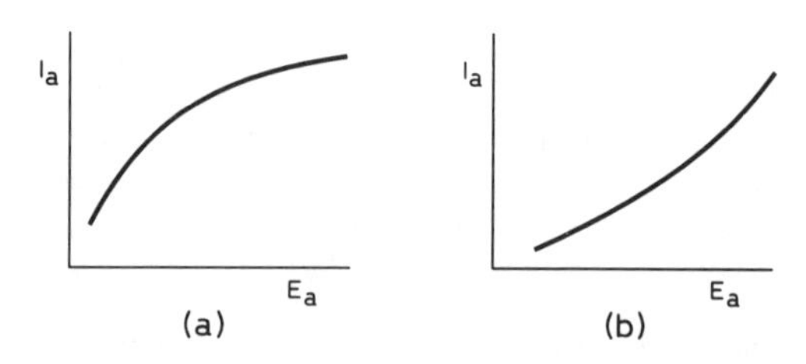

Fig 3.34. Difference between I_a/E_a curves of (a) tetrode and (b) triode

Alternatively the input to an RF amplifier may be applied to the cathode of the valve, with the control grid connected to earth. This is known as 'grounded-grid operation'.

The amplification of an SSB waveform must be achieved with the minimum of distortion, hence the requirement for a linear amplifier.

This requirement is met by a grounded-grid triode amplifier operating in Class AB. Such a amplifier is also suitable for telegraphy.

Important characteristics of valves are:

1. Heater/filament voltage and current.
2. Maximum anode voltage and current.
3. Control grid voltage.
4. Maximum anode dissipation; ie the heat which can be dissipated safely at the anode.

CHAPTER 4

Transmitters

Radio transmitters, irrespective of their application, must:

(a) produce the output power required;
(b) not drift in frequency, ie the frequency shall not change for any reason once it has been set to the desired value;
(c) generally be capable of operating on several frequencies and probably on more than one mode;
(d) have a pure output waveform, ie it should not contain harmonics or any other unwanted frequencies, as may be caused by parasitic oscillations.

A transmitter therefore consists of a 'power amplifier' (PA) to provide the necessary output power which is preceded by low-power drive circuits to generate the RF input to the PA at the required frequencies. The drive circuits may well include the means by which modulation of the transmitter is achieved (see Fig 4.1).

An amateur transmitter is required to operate on any frequency in a number of different bands. As 'netting' (that is, stations in communication operating on the same frequency) is almost universal in amateur radio, it is necessary to be able to set the transmitter frequency to any particular value.

A variable frequency oscillator (VFO) is therefore the basic RF source for the drive circuits.

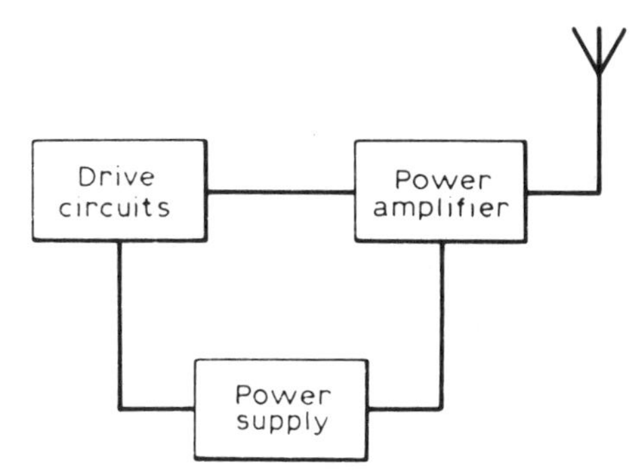

Fig 4.1. Block diagram of a transmitter

Transmitter drive circuits

The transmitter drive circuits may consist of:

(a) a frequency multiplier chain;
(b) a mixer VFO; or
(c) a frequency synthesiser.

The frequency multiplier chain

Table 4.1 lists the limits of the original HF amateur bands, ie up to 30MHz, and the harmonic relationship between these bands.

It will be seen from Table 4.1 that the output of a VFO which is tuneable over a frequency range of 1.75MHz to 2MHz can be multiplied to produce frequencies in all bands up to 28MHz.

It should be noted that any undesired change in the frequency of the VFO is also multiplied in the frequency multiplier circuit; in the worst case by a factor of 16, ie 1.8MHz to 28MHz.

It is particularly important to prevent the radiation of the harmonics other than the required one which are generated in this arrangement.

Harmonics of frequencies in the HF amateur bands will occur around 42MHz, 56MHz and 63MHz. These are liable to cause breakthrough to radio services operating near these frequencies. The frequency multiplier chain cannot be extended to include the WARC bands, ie 10MHz, 18MHz and 24MHz, because these bands are not in harmonic relationship to each other or to the original bands. This circuit is now rarely used commercially, but it provides a simple solution for the home constructor who is interested only in CW operation on the original HF bands.

The mixer-VFO circuit

Coverage of all the HF bands can be achieved by mixing the outputs of a VFO and a fixed-frequency (crystal) oscillator.

Table 4.1. Harmonically related HF bands

Limits of HF bands (kHz)	Equivalent to
1810–2000	1810–2000 × 1
3500–3800	1750–1900 × 2
7000–7100	1750–1775 × 4
14,000–14,350	1750–1794 × 8
21,000–21,450	1750–1787 × 12
28,000–29,700	1750–1856 × 16

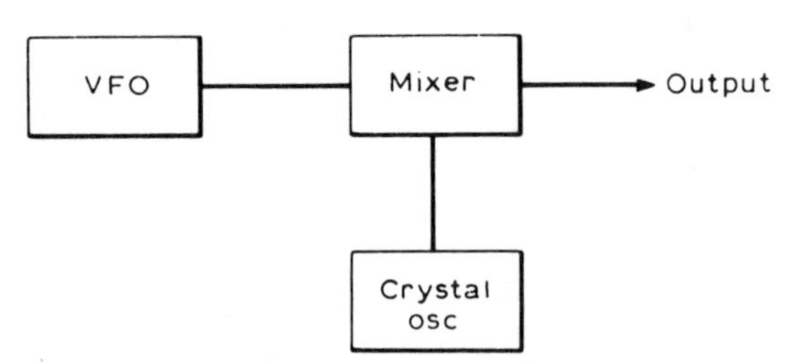

Fig 4.2. Basic block diagram of a mixer-VFO

By the use of a crystal of the appropriate frequency this circuit, generally known as a 'mixer VFO', can produce an output at almost any frequency. Harmonic relationship is not necessary as the desired frequency is the result of the addition of two frequencies and not the multiplication of one (see Fig 4.2).

Typically the frequency coverage of the VFO would be 500kHz in the range of 3MHz to 8MHz. Any unwanted change in the frequency of the VFO remains the same at all output frequencies because it is not multiplied as in the arrangement of Fig 4.2. Thus the overall frequency stability is significantly improved.

For reasons which will be explained later in this chapter, the mixer VFO must be used in an SSB transmitter.

The frequency synthesiser

The term 'frequency synthesiser' has been applied to the mixer VFO just described, but today it normally implies a complex circuit which can produce a large number of equally spaced frequencies, the basic stability of which is determined by a single quartz crystal.

Such a circuit could replace the VFO or the local oscillator as the controlling frequency source in a transmitter and/or receiver in a communication system utilising a large number of equally spaced frequency channels. These systems are widely used in aeronautical, mobile or amateur communication, particularly at VHF and UHF.

The basis of the frequency synthesiser is a 'phase-locked loop' (PLL). This is an electronic feedback loop and is shown in outline in Fig 4.3.

A voltage-controlled oscillator (VCO) is an oscillator whose frequency can be controlled by the variation of a voltage, eg variation of the frequency of an L-C oscillator by a varactor diode. The VCO frequency f_0 is fed back to a phase detector circuit in which it is compared with a reference frequency f_r. The output of the phase detector is an 'error' voltage. This is a varying direct voltage which is proportional to the difference in frequency and phase between f_r and f_0. The error voltage is filtered to remove any high-frequency components from the output of the phase detector and is then fed back to the VCO where it causes the VCO frequency to change in order to reduce the difference between f_r and f_0. This process continues until the two frequencies are equal. The loop is then said to be 'phase locked'.

The phase-locked loop may be used in conjunction with a conventional oscillator to provide a frequency stability which is greater than that of the oscillator itself. However, when it is used with a 'divide-by-*n* counter', it becomes the basis of the frequency synthesiser.

A divide-by-*n* counter is a digital logic integrated circuit which produces a single output pulse for every *n* input pulses; *n* of course is a whole number.

Many counters are available, capable of dividing by fixed ratios according to the way in which they are connected. These can be connected in cascade to give greater ratios, ie ÷10, ÷6 and ÷5 counters in cascade will divide by 300 (10 × 6 × 5). A more useful version is the programmable counter which typically can divide by any number between 1 and 10 according to whether the appropriate pin on the IC is grounded or not.

A basic block diagram of a frequency synthesiser is shown in Fig 4.4. The reference frequency is obtained from a low-frequency crystal oscillator (typically 1MHz to 5MHz) followed by a divider. By selection of the divide ratio of the reference frequency divider and the programmable divider in the feedback loop, a stable and programmable output frequency is obtained. This output frequency cannot be adjusted in a continuous manner as in a VFO but is tuned in a number of discrete steps. In FM equipment it is convenient to make these steps equal to the channel spacing, ie 25 or 12.5kHz. In the case of an SSB transceiver the step needs to be much smaller in order to ensure that the signal can be recovered correctly. 100Hz should be regarded as the maximum step size on SSB unless some other form of fine tune is available.

The actual programming of a frequency synthesiser is achieved in a number of ways, from simple switches to microcomputers. Tuning knobs coupled to electro-mechanical or mechanical-optical switches are also used to simulate conventional tuning. The method used depends upon the complexity and hence price of the equipment. The actual frequency is shown on a digital display.

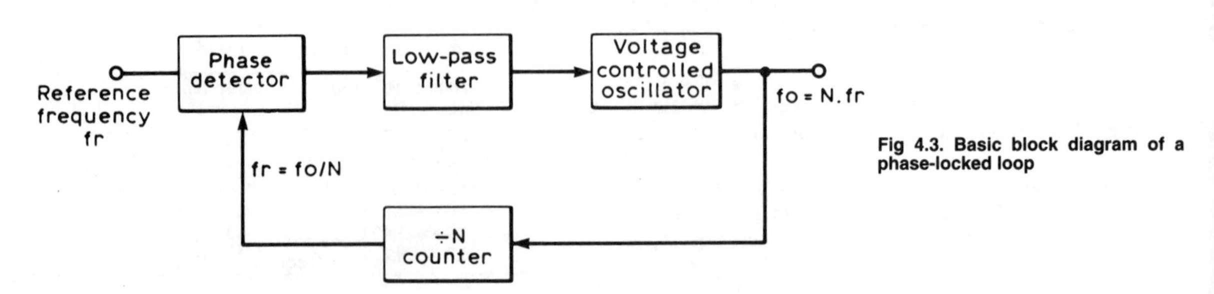

Fig 4.3. Basic block diagram of a phase-locked loop

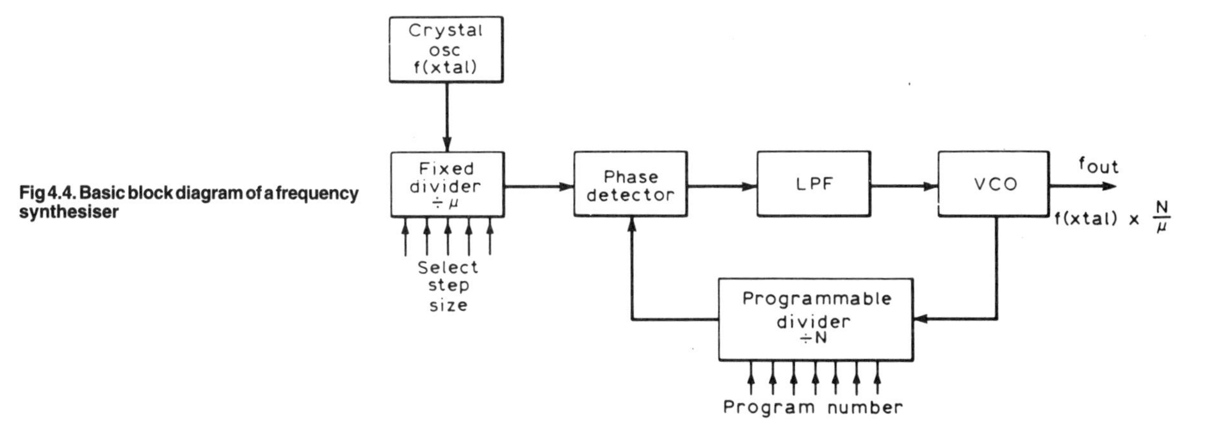

Fig 4.4. Basic block diagram of a frequency synthesiser

Transmitter design

The VFO

Any of the many different oscillator circuits may be used as the VFO, but the generally used circuits are the Colpitts or the Clapp-Gouriet. The factor of prime importance in the design of the VFO is its frequency stability, because:

(a) the frequency must not drift outside the permitted band;
(b) the frequency must not drift outside the pass band of the receiver which is tuned to the transmission.

Frequency stability is particularly important in a transmitter intended for single sideband operation.

The frequency of an oscillator is determined by the resonant frequency of its tuned circuit, thus

$$f = \frac{1}{2\pi\sqrt{LC}}$$

Therefore the magnitude of L, the inductance of the coil, and of C, the capacitance necessary to tune L to the required frequency should not vary under any circumstances.

The factors which may cause L and C to vary are:

(a) mechanical shock:
  (i) The coil must be tightly wound (preferably under tension) on a grooved former, preferably ceramic or made from a low-loss material.
  (ii) The variable capacitor should be of good quality, ie mechanically sound.
  (iii) The wiring between the coil and capacitor should be as short and rigid as possible, so that it cannot move.

(b) change in ambient temperature:

  The coil and capacitor inevitably have a temperature coefficient, the effects of which are manifest as a mechanical movement, hence (a) (i) and (a) (ii) above are doubly important.

Preferably the coil should be air-cored as any form of dust-iron core will cause an additional variation of the inductance of the coil with temperature. The effect of temperature on the tuned circuit may be minimised by locating it as far as possible from any source of heat.

In the Colpitts and similar oscillators, the effect of change in transistor capacitances is minimised by the particular circuit configuration (see Fig 4.5). C1 and C2 are large and swamp the transistor capacitances, but they are part of the frequency-determining capacitance, and so their change in value with temperature is important. These capacitors should be of good quality, preferably having mica or polystyrene as the dielectric. These will have a small positive temperature coefficient which may be compensated for by the experimental addition of a low-value ceramic dielectric capacitor which has a large negative temperature coefficient (C3 in Fig 4.5).

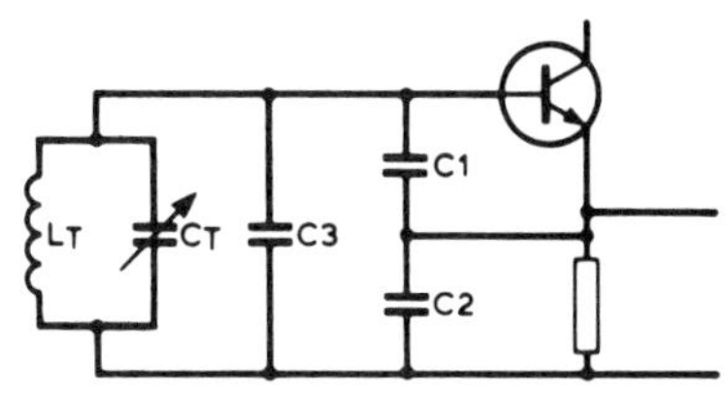

Fig 4.5. Connection between tuned circuit (L_T C_T) and transistor in Colpitts-type oscillator

To summarise, the tuned circuit of a VFO must be soundly made and use best-quality components.

External influences on the VFO

The VFO should be considered as a source of a small output voltage at a constant frequency. It should not be close to any source of heat. The VFO should be lightly loaded by the following stage; excess loading may ultimately stop the oscillation, and any change in loading, if heavy (eg during keying), may well cause a change in frequency ('chirp'). The VFO should ideally be followed by an isolating stage such as a Class A buffer amplifier.

Variation of supply voltage can cause a change in transistor parameters and hence in frequency. Thus stabilisation of

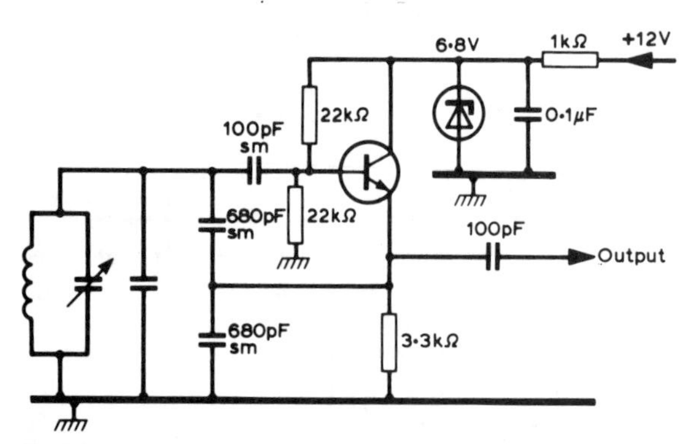

Fig 4.6. Typical Colpitts oscillator circuit

the supply voltage is advisable. This is normally achieved by the use of a zener diode (see Chapter 6).

For frequency-stability reasons, it is considered inadvisable to key a VFO; furthermore it should operate at the lowest possible frequency. This is because the effects of the likely sources of frequency drift increase as the frequency increases. It should be noted that some frequency drift is inevitable during the first few minutes of operation after switch on from cold; it is the drift after this period which must be reduced to the minimum.

Fig 4.6 is the circuit diagram of a Colpitts VFO.

Frequency multiplication

Frequency multipliers are normally low-power transistors operating in Class C in order to produce a collector current which is rich in harmonics. The circuit of such a stage is shown in Fig 4.7.

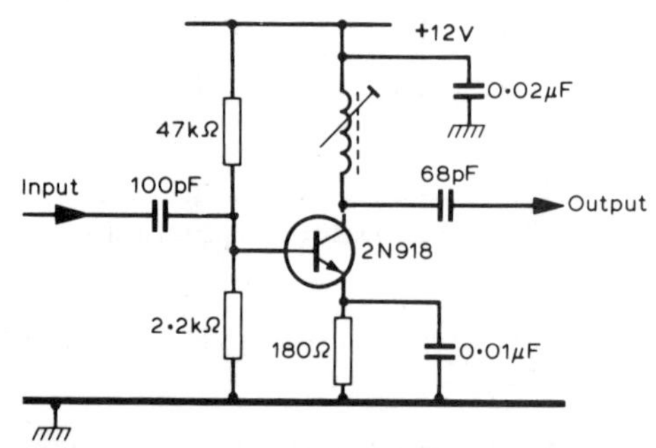

Fig 4.7. Basic circuit of a frequency multiplier

The tuned circuit is resonant at $n \times f$, where f is the frequency of the input voltage and n is usually 2 or 3. Higher multiplications may be used but the output is rather low. The tuned circuit may consist of a small coil and variable capacitor or a slug tuned coil and fixed capacitor.

Design of frequency synthesisers

Synthesisers are notoriously difficult to design and build with a low spurious output level. The design of the phase comparator and loop filter is very critical because any noise or AC components will modulate the VCO and produce unwanted sidebands on the output frequency. These can give rise to considerable interference on transmit. Also there is the possibility of the system going 'out of lock'. This may mean that the VCO could wander completely out of control up and down the band.

Exciters for VHF and UHF transmitters

A stable frequency source followed by a multiplier chain is a convenient arrangement for the VHF (70 and 144MHz) and UHF (430MHz) bands as the transmitters used at these frequencies are generally single-band transmitters. Due to the increased circuit losses at the higher frequencies, a buffer amplifier may be necessary before the output stage.

For stability reasons, crystal oscillators (frequencies in the region of 6MHz, 8MHz or sometimes 12MHz) can be used, but phase-locked loops are usual. A block diagram of an exciter for 144MHz is shown in Fig 4.8.

It will be seen that the addition of a further tripler stage will give an output at 432MHz and that with a slightly different crystal frequency, the first two multiplying stages will provide an output at 70MHz.

Transistor power amplifiers

Transistor RF power amplifiers differ significantly in circuit design from valve amplifiers. This arises mainly because of the much lower input and output impedances of the transistor compared with the thermionic valve. These impedances decrease as the power level increases.

It must be remembered that transistors have very little thermal overload capacity and so the standard precautions for their use are particularly important in high-power amplifiers, ie close attention must be paid to the de-rating factor and heat sinking. Considerable de-rating may be advisable if amplitude modulation is to be used, and therefore SSB or NBFM is to be preferred.

Layout and bypassing of the collector supply must be carefully considered to avoid the creation of instability and hence probably interference (see Chapter 8). The antenna must be accurately matched to the transmitter output as transistor output stages are very sensitive in this respect.

Adjustment and operation are generally more critical.

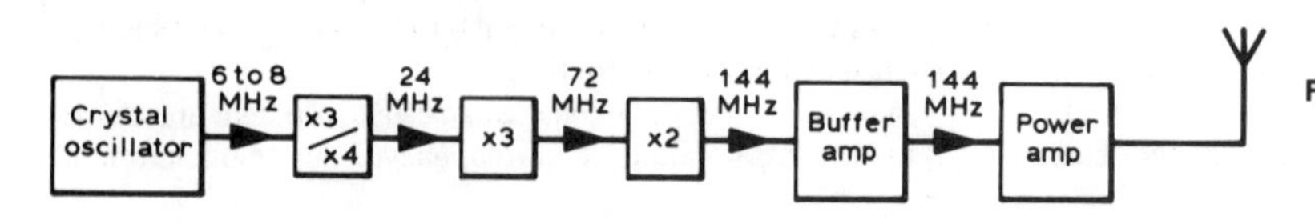

Fig 4.8. Block diagram of 144MHz transmitter

Power amplifiers at VHF

For low powers, ie less than 25W (output) or so, the transistor is a relatively cheap and very convenient solution, particularly for mobile or portable operation when the power level desired can be obtained from a 12V supply.

The generally preferred output arrangement for a transistor PA at 144MHz is the L-pi circuit. This is a combination of the series-tuned L and the conventional pi circuit.

A typical circuit is shown in Fig 4.9. This uses a Mullard BLY83 and gives an output of 7W for a collector supply voltage of 12V, and 12W at 24V. These powers are capable of very good performance at VHF due to the much more effective antennas which are possible at these frequencies. The collector supply is decoupled by 0.01μF and 1,000pF capacitors in parallel. The 0.01μF capacitor next to the output socket is for DC blocking. The relative simplicity of the circuit, which was built on a printed circuit board measuring 127 by 38mm, should be noted.

Power amplifiers at HF

A similar arrangement to that of Fig 4.9 can be used for single-band low-power work on the HF bands, but in general the requirements here tend to be a great deal more severe, ie higher powers are generally called for with operation over all bands.

The lower input and output impedances of transistors which were referred to earlier mean that in a PA having an output of 100W, these impedances will be of the order of 1–10Ω. Conventional pi-network designs at such low impedances lead to impractical values of inductance and capacitance.

The solution adopted at high power is to transform the impedance of the RF input to a transistor PA down from 50Ω and then transform the output impedance of the transistor back up to the normal 50Ω. The basic arrangement is shown in Fig 4.10. The transformers used have ratios of about 4:1 and two such transformers may be used in cascade to provide a higher ratio.

These transformers are known as 'transmission line' or 'broadband' transformers and typically consist of relatively few turns of wire on a toroidal core. The primary and secondary windings are wound together as a pair of wires (a bifilar winding) which may be twisted together.

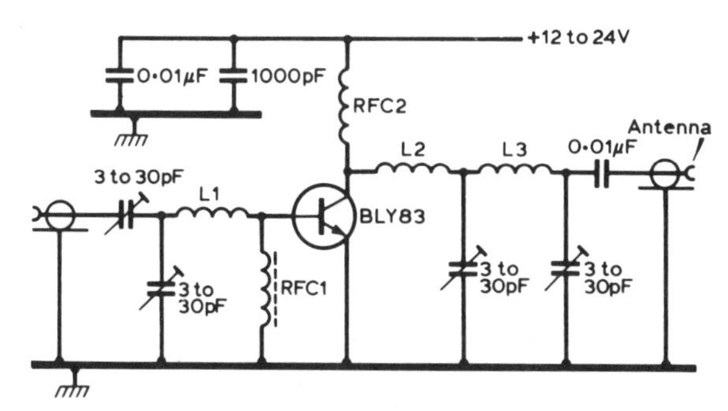

Fig 4.9. Low-power amplifier for 144MHz

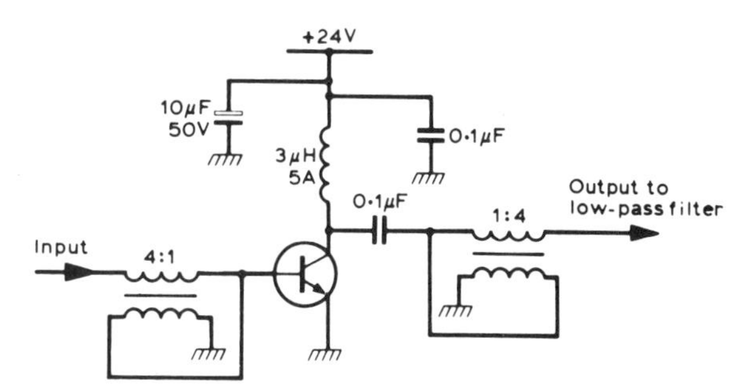

Fig 4.10. Basic circuit diagram of PA suitable for use on the HF bands

Because this type of transformer is broadband, it can operate over a wide frequency range, ie from 3–30MHz. If the drive circuits are similarly designed, the whole transmitter becomes broadband, ie apart from the VFO no band switching or retuning is required. This is obviously highly satisfactory from the operational point of view; however, the design of such transformers is complex and the constructional work involved is not for the inexperienced amateur.

The fact that the RF circuits are broadband means that any harmonic or spurious frequency which may be generated is also amplified and appears at the output. Consequently the output must be filtered to avoid the radiation of any unwanted frequencies. A separate (switched) filter for each band is often used.

Modulation

The output of a transmitter is known as the 'carrier wave'. This is an alternating waveform of the desired amplitude and frequency. Ideally it is a pure sinewave, ie it should not contain harmonics, nor any other unwanted frequencies which might cause breakthrough to other services (TV, broadcast etc). The amplitude of the carrier wave depends upon the power and the output impedance of the transmitter which is most likely to be 50 to 75Ω.

In order that a transmitter may be used to convey a message or other information to a listener, the carrier could be switched on and off (ie keyed) in order to produce the dots and dashes of the Morse code (CW telegraphy). Alternatively, some characteristic of the carrier wave may be varied in sympathy with the message. This process is called 'modulation'.

Modulation may be achieved by the periodic variation of:

(a) amplitude, hence 'amplitude modulation' (AM); or
(b) frequency, hence 'frequency modulation' (FM).

The rate of variation of amplitude or frequency of the carrier wave, ie the 'modulating frequency' (f_m) is assumed to be low compared with the carrier frequency (f_c).

For optimum performance, each mode of modulation requires a particular form of demodulation circuit in the receiver. These are described in Chapter 5.

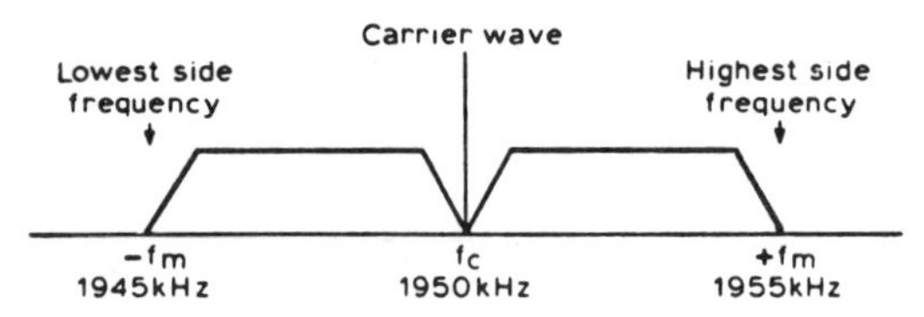

Fig 4.11. Relationship between carrier and sidebands in an AM system

Sidebands

In all modulation processes frequencies above and below the carrier wave are produced, these being termed 'side frequencies'. The bands of side frequencies are called 'sidebands'.

In AM, the highest side frequency is the sum of the carrier frequency (f_c) and the highest modulating frequency (f_m), eg if f_c is 1950kHz and f_m is 5kHz, the highest and lowest side frequencies are 1955kHz and 1945kHz, ie the sidebands extend from 1955kHz to 1945kHz as shown in Fig 4.11. The 'bandwidth' occupied by this transmission is 10kHz.

For FM this situation is much more complex and theoretically the sidebands in an FM system are infinitely wide. The change, which is both positive and negative, in the frequency of the carrier (known as the 'centre frequency') is called the 'deviation'. The deviation is proportional to the amplitude of the modulating signal, so that the limits of the 'swing', ie twice the deviation, are determined by the peaks of the modulating voltage. The rate at which the carrier frequency is deviated is equal to the frequency of the modulating signal, eg if a carrier wave of 7075kHz is modulated by a 3kHz tone of specified amplitude to produce a deviation of 2.5kHz, the carrier frequency will swing between 7072.5 and 7077.5kHz (ie the swing is 5kHz) 3000 times per second. If the amplitude of the 3kHz tone were doubled, the carrier frequency would swing between 7070 and 7080kHz but the rate of variation would still be 3kHz.

The ratio of the deviation to the frequency of the modulating signal is the 'modulating index'. This ratio is obviously not constant, as the deviation depends on the amplitude of the modulating signal. Its limiting value, or the ratio of the maximum deviation to the highest modulating frequency, is called the 'deviation ratio'. In the example quoted earlier, the deviation ratio is 2.5kHz divided by 3kHz or 0.833 for the first given amplitude, and 5.0kHz divided by 3kHz or 1.67 when the amplitude of the modulating signal is doubled.

Bandwidth necessary for communication

For the faithful reproduction of speech and music, it is necessary to transmit frequencies over the whole audible range (ie approximately 20Hz to 16kHz). In a communication system, intelligibility rather than fidelity is of prime importance and in the overcrowded conditions of the present-day amateur bands it is obviously most important to ensure that no transmission occupies a greater bandwidth than is absolutely necessary for intelligible communication.

Experience shows that the intelligible transmission of speech requires that only frequencies of up to 2.5–3kHz need be transmitted. Thus the bandwidth of an AM transmission should not be greater than about 5–6kHz. This restriction of the audio bandwidth is achieved by the use of a low-pass filter having a cut-off frequency of about 2.5kHz in the low-level stages of the modulating circuits.

In FM, as applied to communication, the deviation should be restricted so that the bandwidth occupied is approximately the same as in an AM transmission. This is known as 'narrow-band frequency modulation' (NBFM) and the deviation used is ±2.5kHz or so (as compared with the ±75kHz of the high-fidelity broadcast station).

Depth of modulation

The amplitude-modulated wave is shown graphically in Fig 4.12. Here (a) represents the unmodulated carrier wave of constant amplitude and frequency which, when modulated by the audio-frequency wave (b), acquires a varying amplitude, as shown at (c). This is the modulated carrier wave, and the two curved lines touching the crests of the modulated carrier wave constitute the modulation envelope. The modulation amplitude is represented by either x or y (which in most cases can be assumed to be equal), and the ratio of this to the amplitude of the unmodulated carrier wave is known as the 'modulation factor' or 'modulation depth'. This ratio may also be expressed as a percentage. When the amplitude of the modulating signal is increased, as at (d), the condition (e) is reached, where the negative peak of the modulating signal has reduced the amplitude of the modulated wave to zero, while the positive peak increased the carrier amplitude to twice the unmodulated value. This represents 100% modulation, or a modulation factor of 1.

Further increase of the modulating signal amplitude, as indicated by (f), produces the condition (g), where the carrier wave is reduced to zero for an appreciable period by the negative peaks of the modulating signal. This condition is known as 'over-modulation'. The breaking up of the carrier in this way causes distortion and the introduction of harmonics of the modulating frequencies, which are radiated as spurious sidebands; this causes the transmission to occupy a

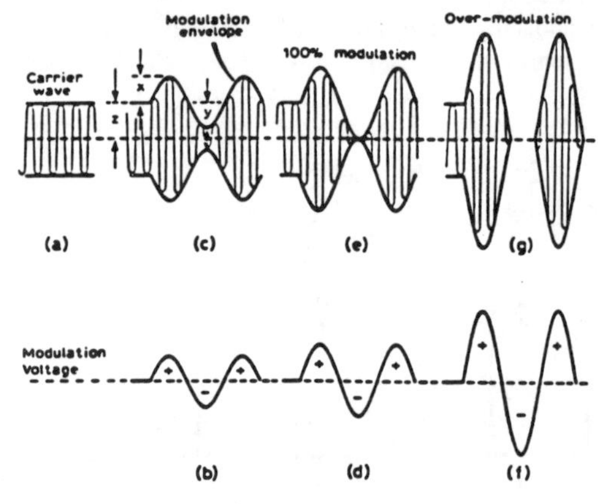

Fig 4.12. Graphical representation of amplitude-modulated wave

much greater bandwidth than necessary, and considerable breakthrough is likely to be experienced in nearby receivers (see Chapter 9). The radiation of such spurious sidebands (sometimes known as 'splatter') must be avoided at all costs.

There is no direct equivalent of over-modulation in an FM system; an increase in the amplitude of the modulating signal will cause an increase in the deviation produced by the transmitter. The recommended deviation would therefore be exceeded and the transmission would occupy a wider bandwidth, ie it would be FM rather than NBFM.

Ultimately the maximum deviation possible is restricted by the design of the RF circuits in the transmitter and the receiver, and attempts to exceed this will result in a distorted signal. However, this would require gross maladjustment and really excessive audio input to the frequency modulator.

Single-sideband operation

Consideration of Fig 4.11 indicates two significant aspects of an amplitude modulation system.

(a) The carrier wave itself does not contain any intelligence, its frequency being f_c.
(b) Both sidebands are identical as they both result from the modulating frequency f_m and the width of each is equal to the highest modulating frequency. Both sidebands therefore carry the same intelligence.

It follows that the carrier need not be transmitted and, as both sidebands contain the same intelligence, only one of them need be transmitted. This has led to the adoption of the system known as 'single sideband suppressed carrier', generally abbreviated to 'SSB'.

Keying

Keying is the switching on and off (by a Morse key) of a transmitter in order to break up the carrier wave into the dots and dashes of the Morse code. Keying implies the switching of an electric circuit and therefore, in order to minimise sparking at the contacts of the key, it should take place at a point in the circuit where the power or current is at a minimum.

In order to avoid the possibility of causing chirp and small changes in transmitter frequency, it is recommended that the VFO itself should not be keyed. The logical point to key is therefore the stage after the VFO, which ideally should be an isolating buffer amplifier, although it may be the first frequency multiplier stage.

The process of keying may cause serious interference. This and the steps to overcome it are discussed in Chapters 8 and 9.

Modulation methods

Amplitude modulation

It can be shown that the effective power in a carrier wave modulated to a depth of 100% by a sinusoidal modulating signal (ie a single pure tone) is 1.5 times the unmodulated carrier power. Thus to fully modulate the carrier, the power in it must be increased by 50%.

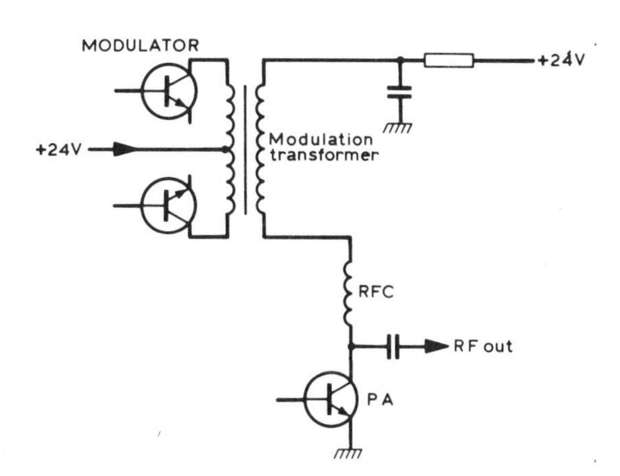

Fig 4.13. Amplitude modulation using transformer coupling to PA stage

This extra power, which is in fact 'added' to the HT supply to the transmitter PA stage, must be supplied by the 'modulator', which is a fairly high-power audio-frequency amplifier. To fully modulate a transmitter operating at an input of 150W to the PA stage would require an AF power of 75W.

The output of the modulator is coupled into the HT supply to the PA by the modulation transformer as shown in outline in Fig 4.13. The ratio of this transformer is determined by the output impedance of the modulator and the modulating impedance of the PA stage.

This is the most effective method of amplitude modulation as the PA stage operates in Class C, giving the highest efficiency. The disadvantage is that large and expensive modulating equipment is necessary for full-power operation, ie as indicated above the basic power required would be 75W, to which must be added an allowance to cover losses in the modulation transformer, so that the design aim should be 100W or so.

Frequency modulation

Frequency modulation is modulation of a carrier wave created by the variation of the carrier frequency by the modulating voltage. The variation (deviation) is positive and negative from the actual carrier frequency (ie the centre frequency).

FM may be achieved by the direct action on the tuned circuit of the fundamental oscillator by a varactor diode. This is shown in outline in Fig 3.18. The diode would be biased so that the variation in capacitance upwards and downwards by the modulating voltage is both equal and linear.

It must be remembered that when the final frequency of an FM transmitter is obtained by multiplication, the deviation of the fundamental is also multiplied.

At VHF, crystal control is common, eg at 144MHz multiplication of ×18 is typical. For a final deviation of 2.5kHz, the fundamental oscillator has to be deviated by less than 200Hz. The frequency of an 8MHz crystal oscillator can be pulled by this small amount by means of a varactor.

Single sideband

The single-sideband (SSB) transmitter performs two distinct functions. These are:

(a) suppression of the carrier wave; and
(b) elimination of one sideband.

The carrier is suppressed by feeding the output of the carrier frequency oscillator and the modulating voltage into a circuit known as a 'balanced modulator'. This is a form of bridge circuit which, when correctly balanced, will cause the RF input to be suppressed, so only the two sidebands appear at the output. The unwanted sideband is removed by a bandpass filter about 2.7kHz wide, shape factor 2–2.5. If for any reason suppression is not 100%, the possibility of some unwanted radiation occurring outside the band must be borne in mind if one is operating right on the band edge.

This combination, shown in block form in Fig 4.14, is called the 'sideband generator'. The single sideband so produced is in fact a band of frequencies corresponding to one sideband of an AM system. It can be produced at a fairly low frequency, typically 455kHz, in which case the frequency band is 455–458kHz. Alternatively the sideband can be generated in the megahertz region, eg at 9MHz, in which case the frequency band would be 9–9.003MHz.

From consideration of a typical band of frequencies, it is clear that multiplication cannot be used to cover several bands. For example, suppose the sideband is generated at 3.5MHz. The frequency band is then 3.5–3.503MHz. If this is doubled it becomes 7–7.006MHz, ie the width of the sideband is also doubled. This is obviously not acceptable and hence frequency translation to different bands must be done by the process of frequency mixing as mentioned in Chapter 2.

Fig 4.15 is a block diagram of a mixer-type exciter in which the outputs of a VFO and a crystal oscillator are mixed to produce an output in the required amateur band.

The VFO will typically cover a range of 500kHz; its actual frequency (order of 3–8MHz) and the crystal frequencies must be chosen so that the wanted and unwanted products in the output of the mixer are as far apart in frequency as possible. The mixer is followed by a tuned amplifier having coupled tuned circuits in the output to improve the rejection of the unwanted product.

This exciter followed by a conventional PA is a preferred,

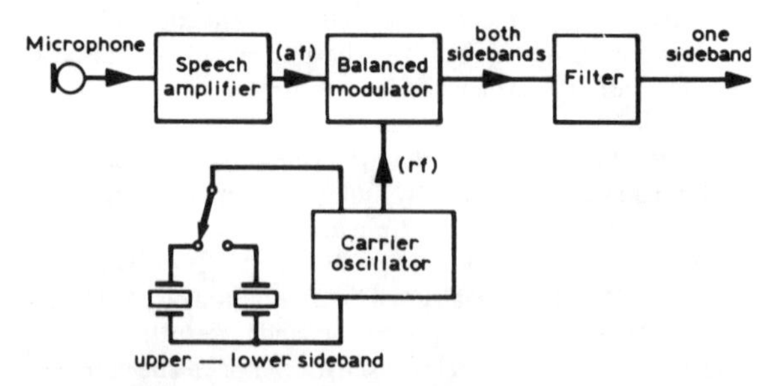

Fig 4.14. Block diagram of SSB generator

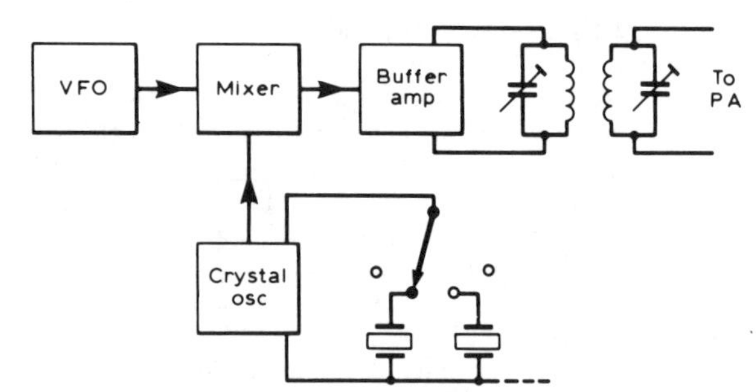

Fig 4.15. Block diagram of mixer-type VFO

although complex, alternative to the frequency multiplier circuit for the following reasons:

(a) the absence of internally generated frequencies which may be radiated and so cause interference to other services; and
(b) the overall frequency stability is improved because it is constant and not dependent on the band in use.

For the reasons discussed earlier it must be used in the SSB transmitter.

Depending upon the frequency at which the sideband is generated, one or more mixing processes may be required to reach the final frequency and to introduce the output of a VFO.

Fig 4.16 is a block diagram of an SSB transmitter in which the SSB is generated at a high frequency, say 9MHz, hence only one mixing process is needed to translate the SSB generation frequency to the output frequency. This diagram is a combination of Figs 4.14 and 4.15.

The PA of the SSB transmitter has to amplify a modulated RF signal without distortion. It therefore operates in Class B rather than Class C and is known as a 'linear amplifier' as the relationship between the output and input is linear.

By convention the lower sideband is transmitted at radio frequencies below 10MHz and the upper sideband above 10MHz. The sideband required is selected by switching the crystal in the sideband generator (see Fig 4.14).

For a number of reasons, the sideband transmitter is combined with the appropriate receiver giving rise to the 'transceiver'. This combination is discussed in Chapter 5.

The modulation systems described above are utilised for the transmission of telephony. The UK licence also permits the use of other more specialised modes of transmission such as radio teletype (RTTY), high-definition television, slow-scan television (SSTV) etc.

These modes are achieved by basically similar forms of modulation. Their use is quite small compared with telegraphy and telephony; they are not included in the RAE syllabus and hence are not covered in this manual.

Transmitter power level

The UK amateur licence now defines maximum power on all bands and modes in terms of output rather than input. The

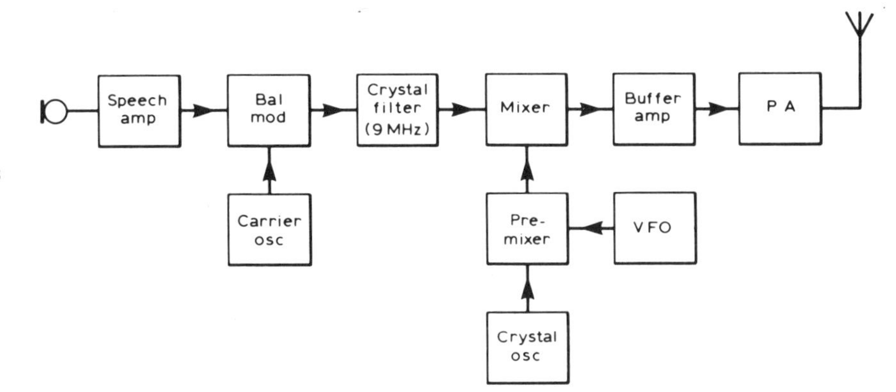

Fig 4.16. Block diagram of SSB transmitter

actual levels are quoted in 'dBW', ie so many decibels above one watt.

Powers in watts and dBW are:

26dBW = 400W	16dBW = 40W
22dBW = 160W	15dBW = 32W
20dBW = 100W	9dBW = 8W

In a CW, AM or FM transmitter it is considerably easier to measure input power than output power. This is because input power is the product of the direct voltage applied to the output stage and the direct current drawn by the output stage. This current does not vary.

The inputs corresponding to the maximum carrier powers now permitted for modes other than SSB (note that there is no change as regards SSB) are shown in Table 4.2. They assume an output stage efficiency of 66 and 55% for output stages operating in Class C and Class B respectively.

Derivation of the SSB power level

In an AM transmitter, the maximum output power permitted (in the absence of modulation) is 100W (20dBW).

When the transmitter is 100% modulated, the amplitude of the peaks of the modulation envelope is twice that of the unmodulated carrier wave (see Fig 4.12). As we are considering a voltage waveform, and since power equals V^2/R (R is the value of the load resistor and V is the unmodulated carrier amplitude), the output power of each cycle of RF energy at the peaks of modulation is $(2V)^2/R$ (because V is the unmodulated carrier amplitude and hence the modulated carrier amplitude is $2V$). This is equal to $4V^2/R$ or four times the unmodulated carrier power. Hence the power at the peak of the modulation envelope or the 'peak envelope power' is 4 × 100 or 400W.

Table 4.2. Inputs and maximum permitted carrier powers

Output	Input Class C	Input Class B
26dBW (400W)	600W	720W
20dBW (100W)	150W	180W
16dBW (40W)	60W	72W
9dBW (8W)	12W	14W

This is the maximum power permitted by the UK amateur licence.

Speech processors

The speech waveform has a high peak but quite low average value. When this is amplified to increase the modulation level, excess deviation, distortion and spreading signal will occur at the peaks. The action of a speech clipper is to limit the peaks. The resulting waveform is filtered as it is now somewhat distorted by the clipping action.

Most modern transmitters incorporate a moderate amount of speech processing which can be switched in and out.

Comparison of modulation methods

AM is now not commonly used; operation is in Class C but efficiency is low due to the audio power required.

FM is widely used for hand-held and mobile operation at VHF and UHF. It gives good coverage over a relatively small area (greater when used via a repeater) but has little use at HF.

SSB is the most effective system from the point of view of bandwidth and power efficiency, and is less adversely affected by disturbances inherent in ionospheric propagation.

Transmitter interference

A high-power transmitter is a potential source of interference to TV, other radio services and domestic AF equipment. All relevant aspects of the installation and antenna system of a transmitter are discussed in Chapters 8 and 9.

The adjustment and tuning of transmitters

Testing and adjustment of transmitters should be done on a dummy load.

The most important adjustment is the audio level. Excess audio will cause over-modulation in all its forms and hence interference to other users of the bands.

An oscilloscope is invaluable for adjustment of an SSB transmitter (see Chapter 10).

Valves as RF power amplifiers

The main use of valves as RF power amplifiers is now as a separate amplifier unit driven by the standard transceiver. The amplifier is generally a 'linear amplifier' operating in the grounded-grid mode as it is required to amplify an SSB waveform. These amplifiers are readily available commercially at ratings well above the UK licence limit. They are quite often home-made as the circuit is quite simple, using relatively few, but some large, components.

Two categories of valve are in use. The first category is the medium-power transmitting valve. Such types with their approximate outputs are: two 572B valves (800–1000W), three 811A valves (600W), two 813 valves (1000W), one 3-5000Z valve (1000W). At VHF and UHF the 4CX250B is used. These valves require a power supply of about 2000V (see Chapter 6).

The other category is colour-TV line output valves ('sweep tubes') with up to four connected in parallel. These are indirectly heated tetrodes used as triodes by connecting both grids together. They have a maximum anode voltage of about 700V (peak E_a is 7kV), with maximum anode current of the order of 350mA. These are not continuous ratings, so the anode dissipation is 25–30W. Such valves are satisfactory at the peaky nature of the SSB waveform. They have been used in commercial equipment and are used in home-built amplifiers. Typical types are 6JS6, 6KD6, 6LQ6 and 6HF6 (American) and PL509 (European, 40V 0.3A heater); they provide an output of about 400W at an anode voltage around 700V.

In the grounded-grid mode it is necessary to isolate the filament of a directly heated valve from ground at RF. This is achieved by supplying the filament via a bifilar choke wound on a ferrite rod as shown in Fig 4.17. There is generally sufficient isolation between cathode and ground at RF if the cathode is indirectly heated. The remainder of the circuit is as that of the standard Class C RF stage. The anode tuned circuit is the standard 'pi' arrangement to attenuate unwanted frequencies.

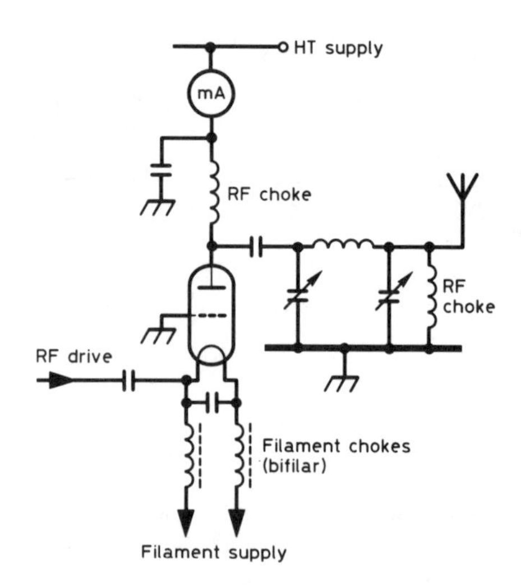

Fig 4.17. Grounded-grid RF power amplifier

The input drive power required by the grounded-grid circuit is somewhat greater than in the grid-driven amplifier. This is offset by the fact that a portion of the drive power appears in the output. The drive power required is within the output capability of most transceivers.

It must always be remembered that an overdriven linear amplifier can create serious interference due to the resulting non-linearity, distortion and excess bandwidth. In other words a linear amplifier should not be used at more than about 75% of its rating.

Few high-power solid-state amplifiers are commercially available; designs for home-built units have been published. No doubt others are available in the application reports of the transistor manufacturers. Problems such as protection against high SWR and power supply failure are likely to be worse at higher powers.

CHAPTER 5

Receivers

Before discussing radio receivers, it is necessary to explain certain terms used to define their characteristics.

'Sensitivity' is the ability of a receiver to receive weak signals. The sensitivity of a receiver is normally quoted as being that level of signal which produces a standard ratio of signal-to-background noise, eg 1μV input to give not less than 10dB signal-to-noise ratio. This means that a 1μV signal will be such that the level of the signal is about three times the level of the noise with the antenna removed and the receiver correctly terminated. The sensitivity of a communication receiver can usually be varied by means of an RF gain control.

'Selectivity' is the ability of a receiver to receive one signal and disregard others on adjacent frequencies. On crowded amateur bands it is necessary for stations to operate on frequencies very close to each other. In order to be able to receive the desired signal with the minimum amount of adjacent-channel interference, a receiver with a high selectivity must be used. Selectivity may be quoted, for example, as follows: 2.5kHz at 6dB down; 4.1kHz at 60dB down.

'Bandwidth'. A receiver with high selectivity is said to have a narrow bandwidth. In the above example, the receiver has a bandwidth of 2.5kHz at 6dB down.

'Frequency stability' is the ability of a receiver to remain tuned to the desired signal. If a receiver is not stable it is said to 'drift'. Stability is determined by the design and construction of the oscillator stage in a receiver.

'Dynamic range' is the range of input signal over which a receiver will function satisfactorily, ie it is the difference in level between the maximum signal the receiver will accept and the minimum signal which will give a usable output. The dynamic range is quoted in decibels (dB); in a good receiver it would be of the order of 90–105dB.

'Automatic gain control' (AGC) is the automatic control of the sensitivity of a superheterodyne receiver by the strength of the signal to which the receiver is tuned. For weak signals, the sensitivity needs to be high, but for strong signals a low sensitivity suffices. AGC is useful when there is fading on a signal; the sensitivity is varied in accordance with the signal to produce an almost constant audio output level.

The superheterodyne receiver

The basic limitations of inadequate selectivity and lack of gain of the early receivers led to the development of the 'supersonic heterodyne' principle in around 1920. In the supersonic heterodyne (or 'superhet' as it is colloquially known) receiver, the frequency of all incoming signals is changed to a fixed, fairly low frequency at which most of the gain and the selectivity of the receiver is obtained.

Because this fixed frequency is lower than the signal frequency but higher than the audio frequency, it became known in the early days of the superhet as the 'intermediate frequency' (IF). As the circuits operating at the intermediate frequency, once adjusted, need no further tuning, high amplification and good stability are possible.

In order to convert the signal frequency to the intermediate frequency, a frequency-mixing process is necessary. In the mixer the signal frequency is mixed with the output of an oscillator, the frequency of which is varied by the receiver tuning control. This oscillator is called the 'local oscillator'.

The resulting intermediate frequency is amplified and fed to detector and audio amplifier stages. The output of the IF amplifier is used to provide a voltage, the amplitude of which is proportional to the amplitude of the input signal. This is used to control the gain of the receiver, giving 'automatic gain control' (AGC), to compensate for variation of the received signal.

In order to receive telegraphy (CW) signals, it is necessary to provide a signal to beat with the intermediate frequency to produce a beat note which is audible. This signal is generated by the 'beat frequency oscillator' (BFO) which operates at the same frequency as the IF but is variable about this frequency by about ±3kHz.

Fig 5.1 is a block diagram of the simplest possible superhet receiver. The basic design implications of each stage of a superhet receiver will now be discussed in greater detail.

Mixers

The mixing process of the superhet receiver is shown in Fig 5.2. As is inevitable in the mixing process, two frequencies appear at the mixer output, these being the sum and difference of the signal and local oscillator frequencies. Only one of these is wanted as the intermediate frequency, and in fact only one frequency is accepted by the following IF amplifier. The reason for this will be seen later.

To take a simple numerical example, assume an IF of

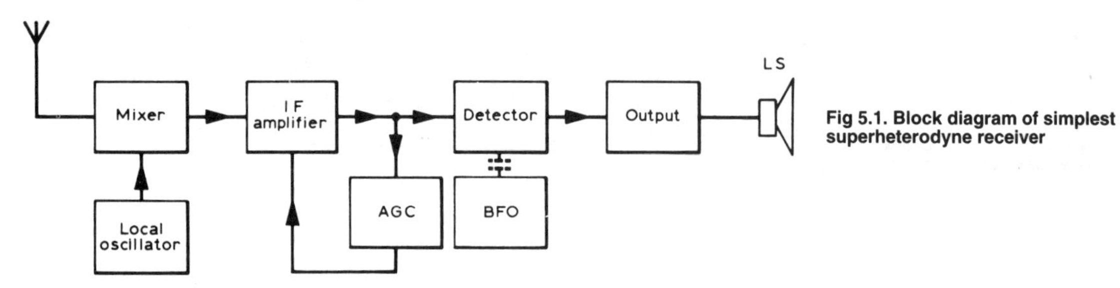

Fig 5.1. Block diagram of simplest superheterodyne receiver

500kHz; if the signal frequency is 1000kHz, the frequency of the local oscillator must be 1500kHz; see Fig 5.3(a).

$$\text{IF} = f_0 - f_s$$
$$500\text{kHz} = 1500\text{kHz} - 1000\text{kHz}$$

However, a strong signal on a frequency of 2000kHz, that is, a frequency which is twice the IF (ie 2 × 500kHz) away from the first frequency (1000kHz), can also produce an intermediate frequency of 500kHz; see Fig 5.3(b). Hence

$$\text{IF} = f_s - f_0$$
$$500\text{kHz} = 2000\text{kHz} - 1500\text{kHz}$$

Thus two signals, the wanted one on 1000kHz and an unwanted one on 2000kHz, can both result in an intermediate frequency of 500kHz. The unwanted signal (on 2000kHz) is called the 'second channel' or 'image'.

This phenomenon is manifest as the reception of two signals apparently on the same frequency and at the same time. The reception of the unwanted signal is known as 'second-channel' (or 'image') interference and it is a 'spurious response' of the receiver in question.

Second-channel interference can only occur when there is a signal on the second channel which is strong enough to reach the mixer. The most common example is the reception in the 20m amateur band of some of the powerful 19m broadcast stations on a simple all-wave receiver having an IF of 455 to 465kHz (the frequency separation between parts of the 20m amateur and 19m broadcast bands is about 920kHz).

It is most likely to occur when the mixer is inadequately screened and the receiver antenna fed direct to the mixer, so there is no RF stage; in other words there is insufficient selectivity at the signal frequency to reject the second-channel frequency.

As the existence of second channel interference depends on the response of the signal input circuit of the mixer to a frequency which is separated from the resonant frequency of the input circuit by twice the IF, it is clear that increasing the IF will reduce the incidence of image interference.

As a result of the selectivity of the tuned circuits therein, one or two RF stages between the antenna and the mixer will also provide considerable attenuation of the second channel.

Fig 5.4 shows a typical mixer/oscillator arrangement using transistors. Bipolar transistors, FETs and MOSFETs are all suitable for this application, and the typical arrangement shown here is a Colpitts oscillator, the collector supply being stabilised by a 6.8V zener diode. The oscillator output is fed to a buffer stage to provide isolation between the oscillator and the mixer. The mixer uses a dual-gate MOSFET which is particularly suitable for mixer applications, having two gates. The input (RF) tuned circuit is between gate 1 and earth, and the output (IF) tuned circuit is between the drain and the 12V supply. The oscillator voltage is applied to gate 2.

The receiver local oscillator has the same requirements of frequency stability etc as the VFO in a transmitter. The discussion of VFO stability in Chapter 4 is therefore equally applicable to receiver local oscillators. The situation is complicated by the necessity for the receiver local oscillator to be switched to cover a number of frequency bands.

Generally the local oscillator frequency is on the high side of the signal frequency. The reason for this is as follows.

Assume a receiver tunes to signals in the range 1500kHz to 4500kHz and has an IF of 1000kHz. The signal frequency range has a ratio of 3 to 1 (4500 to 1500) and because $f \propto 1/\sqrt{C}$, the change in capacitance must therefore have a ratio of 9 to 1, say, 20pF to 180pF.

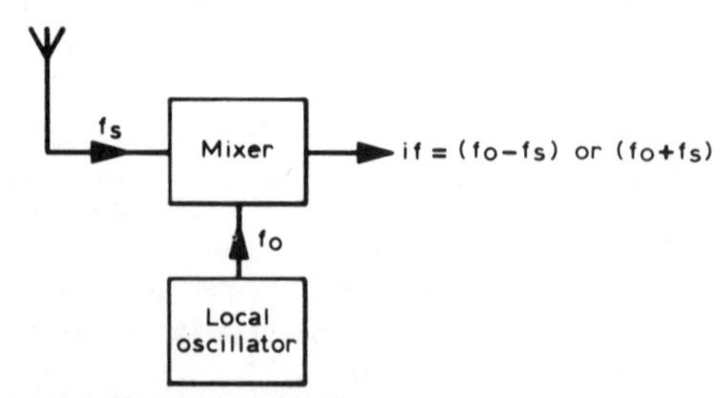

Fig 5.2. Superheterodyne mixing process

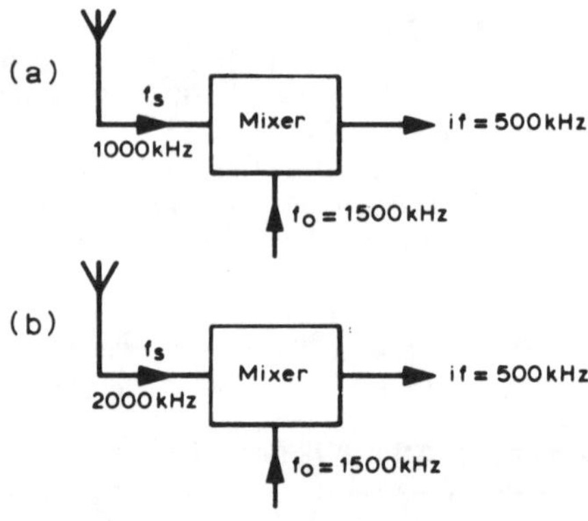

Fig 5.3. (a) Wanted signal. (b) Image frequency

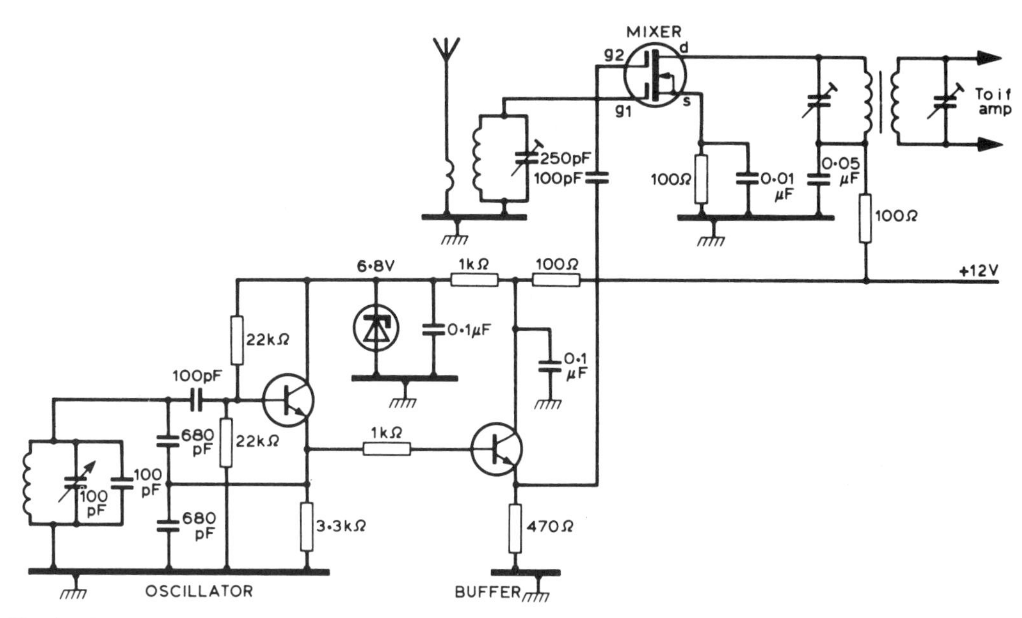

Fig 5.4. Mixer/oscillator circuit

If on the low side, the oscillator would have to tune from 500kHz to 3500kHz, a frequency range of 7 to 1, requiring a capacitance range of 49 to 1, say 20pF to 980pF.

Alternatively, if the oscillator is on the high side it would need to tune from 2500kHz to 5500kHz, a frequency range of 2.2 to 1, requiring a capacitance range of only 4.8 to 1, say 20pF to 96pF. A capacitor having a range of 20pF to 96pF is obviously very much more practical than one which has a range of 20pF to 980pF.

Tracking

The tuned circuit of the local oscillator must maintain throughout its tuning range a constant frequency separation equal to the IF from the mixer tuned circuit. This requisite is known as 'tracking'.

The need for tracking arises because the oscillator and mixer tuned circuits cannot be identical in inductance and capacitance. For example, for a signal frequency range of 5–10MHz with an IF of 500kHz, the mixer tuned circuit must cover 10MHz to 5MHz (ratio 2:1), while if the oscillator is on the high side, its tuned circuit must cover 10.5MHz to 5.5MHz (ratio 1.9:1). Thus the oscillator tuning capacitor often has a smaller capacitance than the mixer capacitor.

The wider the frequency range, the more difficult tracking becomes; in practice the optimum solution generally considered is that tracking should be correct at both ends of the tuning range and also at a point near the middle.

Tracking is generally achieved in the better class of receiver by the careful adjustment of a small trimming capacitor in parallel with the oscillator tuning capacitor at the high-frequency end of each range, and the inductance of the tuning coil (by means of a dust core) at the low frequency end.

RF amplifiers

RF amplifiers, ie tuned amplifiers operating at the signal frequency, are employed in the majority of high-quality receivers and also in transceivers. The tuning is ganged with the mixer/local oscillator tuning control.

Basically, an RF stage improves the sensitivity of the receiver, ie it increases the signal/noise ratio. The additional selectivity resulting from the extra tuned circuits may be advantageous in a number of ways, ie the chance of second-channel interference is reduced, as is radiation from the local oscillator via the antenna. This additional RF selectivity is always useful.

The older receivers with an IF of around 465kHz always employed two RF stages and the second-channel interference then only became unacceptable above about 30MHz. If the IF was 1.6MHz, one RF stage could be considered to be adequate.

The intermediate-frequency amplifier

The function of the intermediate-frequency (IF) amplifier is to amplify the output of the mixer before demodulation; it is a tuned amplifier, ie it operates at a single fairly low frequency (the IF). Hence high gain and stability are easily achieved, in fact it is the IF amplifier which provides virtually all the selectivity and most of the gain of the superhet receiver. Its importance is therefore obvious.

The selectivity of the IF amplifier can be achieved by means of tuned circuits or bandpass filters. It is desirable to be able to change the bandwidth of the IF amplifier to suit the signal being received, ie from about 2.7kHz ('wide') for SSB to about 300Hz ('narrow') for CW.

The tuned circuits are designed as 'coupled pairs' (see

Chapter 2). An IF transformer is such a pair in a screening can (see Fig 5.5). A typical IF amplifier may consist of two or three such stages in cascade.

The value of the IF will depend upon the selectivity required and the need to minimise image interference. These requirements are incompatible, ie low image interference requires a high IF whereas high selectivity requires a low IF (see Chapter 2). One solution to this problem is the 'double superhet', having two different intermediate frequencies. The first is fairly high, typically 1.6 to 3MHz for good image performance; this is then converted by means of a second mixer and local oscillator to a second IF which is low to provide high selectivity. The second IF may be as low as 50 to 100kHz.

Fig 5.6 shows the typical selectivity of an IF amplifier based on tuned circuits. The achievement of wide and narrow bandwidths in such an amplifier presents difficult electrical and mechanical design problems.

A better IF selectivity characteristic can be obtained by the use of a 'bandpass filter'. One such version is based on the use of quartz crystals; two, three or four pairs of crystals carefully matched in frequency may be used. The older general-purpose receiver used a crystal filter which employed a single crystal in conjuction with a phasing capacitor. This simple arrangement gave a nose bandwidth of less than 0.5kHz and so was very useful for receiving telegraphy. However, the skirt bandwidth, being determined mainly by the tuned circuits, was often very wide.

Another form of bandpass filter is the mechanical type. This is a mechanically resonant device which receives electrical energy, converts it into a mechanical vibration which is then converted back into electrical energy at the output. The mechanical vibration is set up in a series of six to nine metal discs by the magnetostrictive effect.

Filters of the bandpass type have a much flatter top to the selectivity curve and shape factors of 1.5–2.5. They are made

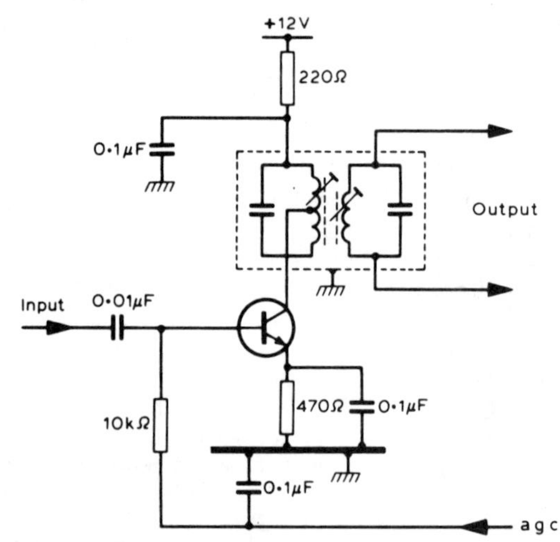

Fig 5.5. IF amplifier circuit

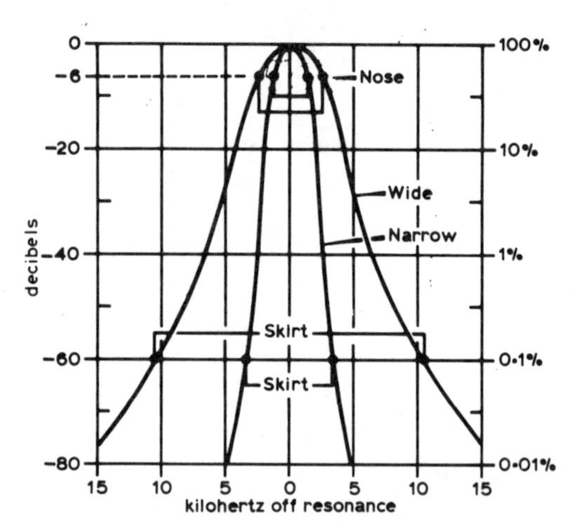

Fig 5.6. Typical overall selectivity of IF amplifier based on tuned circuits

in various bandwidths from 0.3–10kHz at frequencies of commonly 455kHz and 3–9MHz. Such filters of the desired bandwidth can be switched into an IF amplifier, and the design of the IF transformers used then becomes relatively unimportant as the overall selectivity is determined by the filters. They are compact but tend to be expensive. This type of filter has a sufficiently steep characteristic (ie low shape factor) to filter out the unwanted sideband of a double-sideband signal and is therefore the basis of the filter method of single-sideband generation. It is also used in the receive function of the SSB transceiver.

The detector

The purpose of the detector is to rectify or demodulate the output of the IF amplifier, in order that the modulation originally superimposed upon the carrier wave at the transmitter can be recovered as a varying direct voltage, which can be amplified and converted into sound by the loudspeaker.

For optimum performance, each mode of modulation requires that a particular demodulating circuit is used in the receiver.

If the carrier is unmodulated, as in telegraphy, it is necessary to mix with the IF amplifier output another signal of a slightly different frequency which is generated by a 'beat frequency oscillator' (BFO) in order to produce a difference frequency in the audible range, ie an audible beat note which is then recovered by the detector.

Diode or envelope detector

The simplest and most commonly used detector is a single diode operating as a half-wave rectifier as shown in Fig 5.7. The output is developed across the resistor (the diode load) and then fed to the following audio amplifier. This arrangement is also known as an 'envelope detector' as its object is to recover the modulation envelope. It is the normally used circuit for the detection of an amplitude-modulated signal and, in conjunction with a BFO, for CW telegraphy.

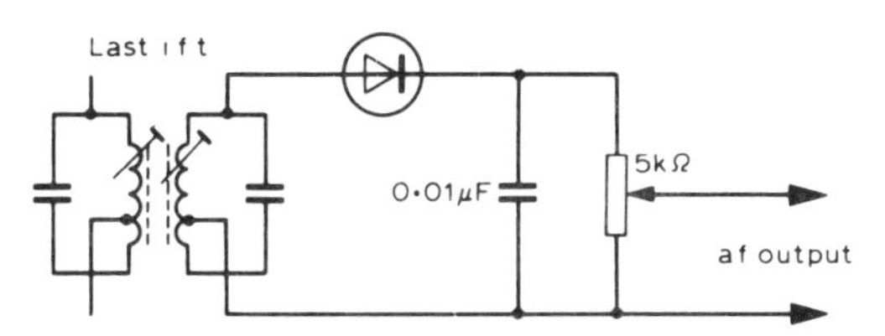

Fig 5.7. Basic circuit of diode detector

Detection of an FM signal

The ratio detector is a circuit which has been widely used in both entertainment and amateur receivers. The basic circuit is shown in Fig 5.8.

Because the primary voltage is injected into the centre tap of L2 by L3, the voltages at the ends of L2 depend on the phase difference between L1 and L3. At resonance, at the centre frequency, the voltages applied to the diodes are equal. As the frequency increases, the voltage at one end increases and decreases at the other end. The reverse occurs when the frequency decreases from the centre value. The voltages at the ends of L2 are rectified by the diodes and so the rectified voltages appearing across C1 and C2 vary, giving rise to an output at point A.

The output is therefore proportional to the ratio of the voltages which appear across C1 and C2. The total voltage across C1 and C2 is held constant (ie its amplitude does not vary) by the capacitor C3, which is at least 8μF.

Detection of an SSB signal

The detection of a single-sideband (SSB) signal necessitates the insertion of a signal into the detector to simulate the carrier wave which was suppressed in the transmitter. This signal is generated in the receiver by the 'carrier insertion oscillator' (CIO).

This function can be fulfilled by the BFO of an AM/CW receiver, and by the use of the usual diode envelope detector, reasonably satisfactory results may be obtained. However, the diode detector system has directly opposite requirements for optimum CW detection and optimum SSB detection. A very small input signal from the BFO is preferable for CW whereas SSB detection requires a much larger BFO signal. As the BFO injection voltage is never adjustable, it should be set to suit whichever mode is of most interest. A large BFO voltage is likely to affect the operation of the AGC system as discussed later in this chapter.

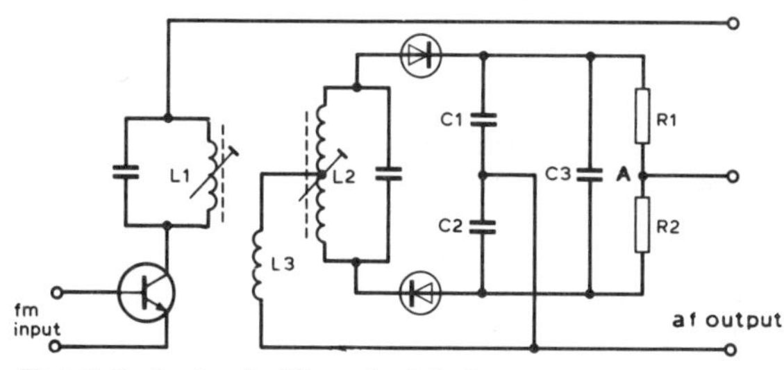

Fig 5.8. Basic circuit of the ratio detector

The 'product detector' is the preferred circuit arrangement for the resolution of an SSB signal. This is a mixer circuit, and one of several varieties is shown in Fig 5.9. The frequencies involved in this mixing process are the receiver IF and the frequency to which the BFO (or CIO) is set. The BFO frequency will have been adjusted to produce an acceptable audible beat frequency. The mixer output frequencies are the audio frequency required (difference) and the sum of the intermediate and BFO frequencies – the following stages will not operate at the sum frequency. The circuit shown uses an FET. The IF amplifier output is connected to the gate and the BFO/CIO injection voltage is taken to the source.

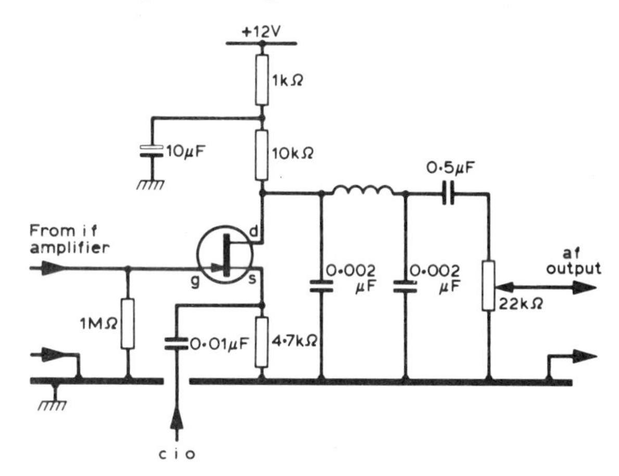

Fig 5.9. Circuit of product detector

The product detector is also a very effective demodulator of CW telegraphy signals. A further advantage is that the BFO injection voltage necessary is small and is the same for SSB and CW.

Thus the modern all-mode receiver will include two detectors, ie a diode envelope detector for AM and a product detector for SSB/CW. The VHF receiver is most likely to also include a ratio detector for FM.

Beat frequency oscillator

The BFO is a conventional oscillator which operates at the IF of the receiver. Its frequency is generally variable by ±3–4kHz by means of a front-panel control in order to provide the audible beat note discussed earlier, and to enable this note to be set at a frequency which is acceptable to the operator. The BFO is switched on and off by a front-panel control. Coupling between the BFO and detector is very loose; 5pF or so, or even by stray capacitance.

Carrier insertion oscillator

The carrier insertion oscillator generates a signal to simulate the carrier wave which has been suppressed in the transmitter. It also performs the same function as the BFO when receiving telegraphy.

In order to achieve the frequency stability necessary in an SSB system, the CIO would be crystal controlled, a separate crystal being used for each sideband. The crystal frequencies are typically ±1.5kHz from the intermediate frequency.

Automatic gain control

Automatic gain control (AGC) refers to the control of the gain of the receiver in sympathy with the strength of the received signal. The object is to ensure that the output of the receiver remains constant or nearly so, irrespective of the incoming signal strength which may undergo considerable variation due to propagation conditions (fading) or simply due to the relative signal strengths of the several stations operating in a net.

The basis of the operation of an AGC system is as follows. As the received signal strength increases, so does the receiver output, and a sample of this is taken from some point in the output stages and fed back in such a way as to reduce the overall gain of the receiver. As the signal fades or a weaker signal is being received, the output falls and a lower control voltage results, hence increasing the receiver gain.

The gain of the IF amplifier is controlled by feeding the AGC voltage to the base of each transistor in order to vary the emitter current and hence the gain.

The point in the receiver from which the AGC control voltage is taken depends mainly on its complexity, and inevitably it is after the IF amplifier. In a simple receiver the AGC voltage would be taken from the detector diode circuit. A separate diode to develop the AGC voltage, fed via a small capacitor (say, 33pF) from the same point as the detector diode, provides a more flexible arrangement from the design aspect and is generally to be preferred.

The simplest arrangement is shown in Fig 5.10. R1–R4 form a divider which provides the diode with a small forward bias to make it more sensitive to weak signals. Any increase or decrease in voltage at point A due to changing signal strength will be applied to the base of the first IF amplifier transistor. Any audio component is filtered out by R2 and C1.

Effective AGC for CW reception presents a number of difficulties. The rectified BFO voltage may well reduce the gain even in the absence of a signal. For this reason AGC is often switched out of operation (by S1 in Fig 5.10, ganged to the BFO on/off switch) when receiving CW signals.

A receiver intended for CW/SSB reception will invariably employ a product detector. This provides much better isolation between the locally generated BFO/CIO voltage and the AGC circuit, and hence AGC on CW reception is much more effective. The AGC voltage in an SSB receiver is sometimes obtained by sampling and rectifying the audio at some point in the audio amplifier. There is not much to choose between the two methods.

By suitable design, an AGC system can provide a characteristic which exhibits little change in output level (less than 4dB) for a very large change in input signal (90–100dB). However, a more significant characteristic, particularly for SSB with the intermittent nature of the signal and its syllabic variations, is the speed of operation of the AGC system. The AGC must take effect quickly: the attack time must be of the order of 2ms but the release should be much slower, about 200–300ms. These times are governed by the time constants (ie products of resistance and capacitance) in the AGC circuit.

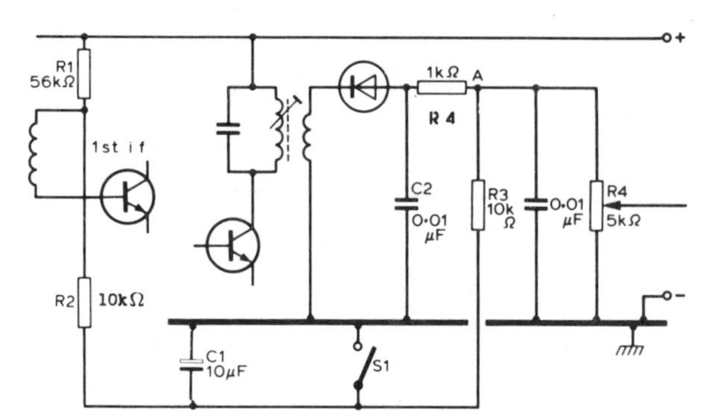

Fig 5.10. Basic configuration of AGC system

Audio stages

The audio side of the communication receiver is conventional in every way, bearing in mind the restricted audio bandwidth necessary for communication purposes. The audio power output is normally 1–2W peak to a small loudspeaker located within the receiver cabinet. Generally provision is made to plug in a pair of headphones at the input side of the output stage.

The inclusion of some form of additional selectivity or filter in the audio chain is not uncommon in the more complex receiver, particularly in the older general-purpose receiver.

This may take two forms, one being a sharp notch filter, in which the notch can be tuned across the audio band. The gain in the notch is very much reduced and so it can attenuate a particular interfering frequency. Alternatively a sharp peak of amplification at a particular frequency, say 1000Hz, may be provided and, by careful adjustment of the BFO to give a 1000Hz beat note, the overall selectivity for CW may be improved.

Calibration oscillator

This is a crystal oscillator arranged to produce a high level of harmonics and operating usually at 100kHz. It provides a calibration 'pip' every 100kHz throughout the receiver tuning range. Generally provision is made to adjust the tuning scale or the pointer slightly to enable the calibration to be corrected at the 100kHz points. In the better class of equipment there are facilities for checking the accuracy of the crystal frequency against a standard frequency transmission.

Noise limiters and noise blankers

Much electrical interference to reception arises from the short pulse of energy radiated whenever a spark occurs, be it from a faulty switch or a car ignition system. The noise limiter is an arrangement of diodes which clip off those interfering pulses which exceed the modulation level in amplitude. The level at which clipping occurs is normally adjustable. The noise limiter is a simple and quite effective circuit.

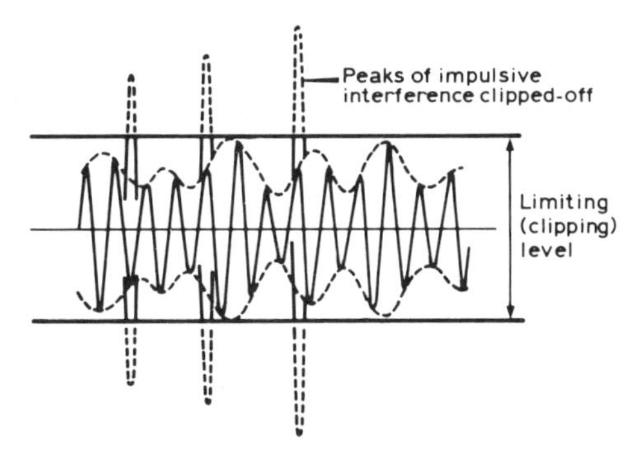

Fig 5.11. Interfering noise pulses on the modulated waveform

Fig 5.11 shows the wanted signal and the unwanted interfering waveform.

The noise blanker is a much more complex circuit in which the interfering noise pulses are selected, amplified and detected. The resulting waveform is then fed back into the receiver via a gate circuit. Thus the interfering pulse waveform is 'blanked' out before it reaches the output stage of the receiver.

Squelch circuits

'Squelch' is the name given to a facility which is normally part of an FM communication receiver (or transceiver). The object is to switch off the audio output of the receiver in the absence of a signal or when the incoming signal strength is inadequate for satisfactory communication, ie when the receiver is just at the maximum range of a particular transmitter. Thus the annoying hiss produced by the receiver in the absence of a signal is eliminated. The level at which the squelch circuit comes into operation is normally adjustable.

Signal-strength meters

Most commercial receivers now incorporate a signal-strength meter (S-meter). Normally this consists of a sensitive milliammeter, often in a bridge circuit, which is used to monitor the AGC control voltage. This of course varies in sympathy with the incoming signal.

The meter is calibrated in S-units up to S9 and decibels up to 40 or 60 above S9. There is no generally agreed definition of an S-unit (it may be 4 or 6dB) or of the zero point of the meter. Unless an S-meter has been specially calibrated against a signal generator on each band, no great reliance should be placed on its readings.

A complete receiver

A block diagram of a double-superhet receiver incorporating the points discussed so far in this chapter is shown in Fig 5.12. For obvious reasons this is known as a 'communication receiver' – it is intended for communication purposes and not entertainment.

Consideration of Fig 5.12 raises the query as to which local oscillator is tuned. In fact, either can be tuned as follows:

(a) LO1 tuned and LO2 fixed frequency;
(b) LO1 fixed frequency and LO2 tuned.

For stability reasons, the fixed-frequency oscillator should be crystal controlled.

The arrangement (a) allows, by appropriate design of the RF/mixer/local oscillator tuned circuits, each amateur band to be spread over the whole of the tuning scale which is obviously a very convenient arrangement.

In arrangement (b) the RF/mixer/local oscillator tuned circuits are designed to cover a small range, usually 500kHz. Each amateur band apart from 28MHz can be covered in one such range; for full coverage of the 28MHz band four 500kHz segments are required.

The second local oscillator frequency (crystal controlled) is chosen to convert the above RF ranges to the IF range required, which may be for example 5000–5500kHz. The rest of the receiver is therefore a single-superhet having a single tuning range of 5000–5500kHz.

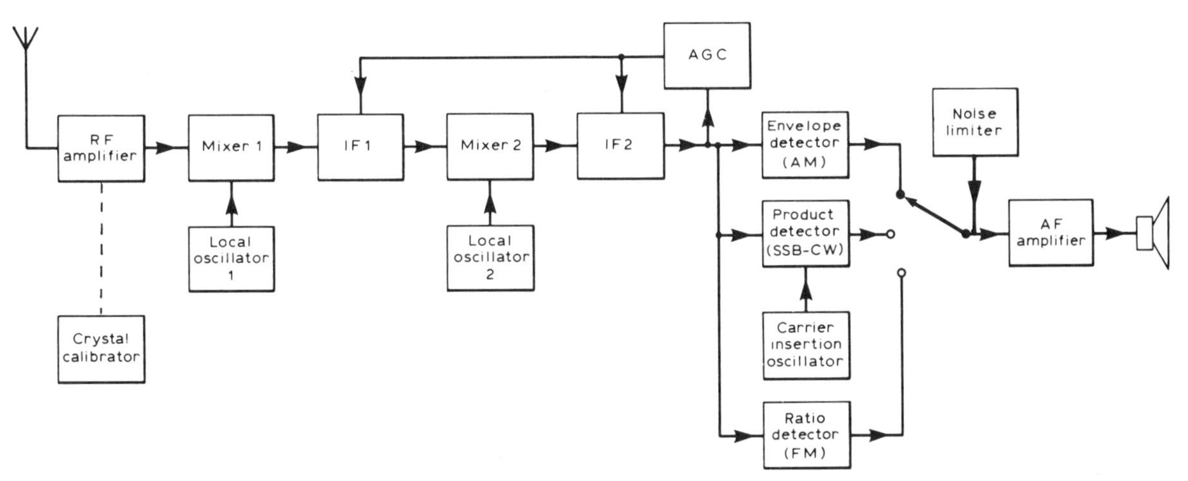

Fig 5.12. Block diagram of double superheterodyne

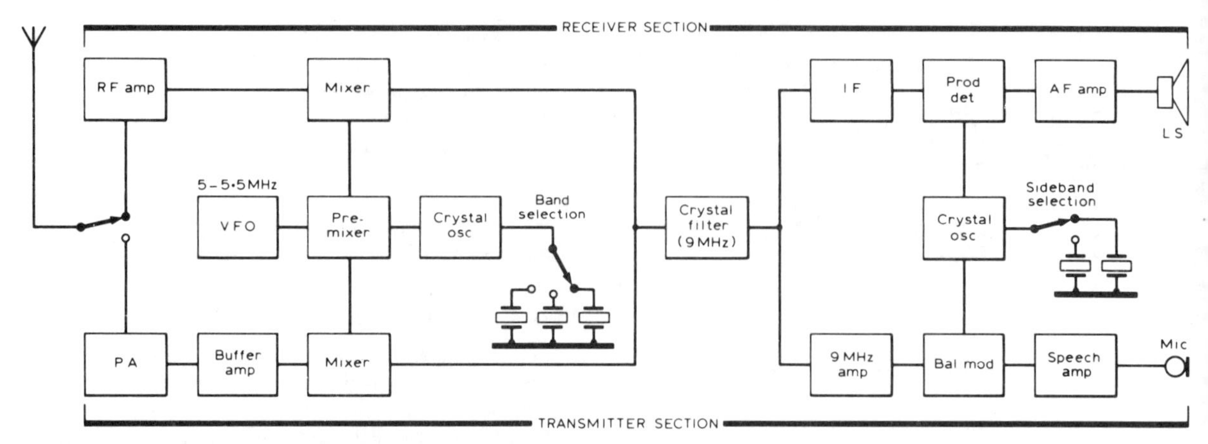

Fig 5.13. Block diagram of SSB transceiver (from *A Guide to Amateur Radio*)

The intermediate frequency would be typically 455kHz and the IF amplifier would ideally contain crystal filters having bandwidths of about 0.4kHz, 3kHz and 6kHz which would be switched in to suit the mode in use.

This is a very common arrangement – it provides a constant tuning rate on each band which is slow enough for satisfactory tuning of an SSB signal, ie a tuning rate of no more than 25kHz per revolution of the tuning knob.

The transceiver

The frequency stability of the transmitter and receiver are especially important in an SSB communication system. The carrier re-insertion process also must be done accurately. Unless the local signal is within 25Hz of the original (suppressed) carrier frequency there will be distortion, and if the frequencies differ by much more than that figure, the SSB transmission will be unreadable.

A narrow band filter (2.5–3kHz) is required to remove one sideband in the transmitter and a similar filter is advantageous in the receiver. Both require a very stable VFO.

For these reasons, the transmitter and receiver circuits can be combined to produce the 'transceiver' in which the same VFO and filter are used in both the transmit and receive functions. This is now the preferred arrangement for SSB operation; it can also be used for telegraphy.

The modern transceiver is compact and inevitably complex in design. It is often followed by a separate linear amplifier to boost the power to the legal maximum. Fig 5.13 is a block diagram of a typical transceiver.

Transverters

The transverter enables an HF bands transceiver to operate on bands other than those for which it was designed. It combines the function of transmit frequency conversion (generally upwards) and receive frequency conversion (generally downwards) using a common oscillator.

The most common application is to provide operation on the VHF and UHF bands, but they have also been designed to transvert to 1.8MHz. Fig 5.14 is a block diagram of an HF bands transceiver (tuned over the 28MHz band) transverted to 144MHz. The transverter oscillator frequency is 116MHz.

VHF receivers

The principles discussed in this chapter apply to receivers for the VHF and higher frequency ranges, although somewhat higher values of intermediate frequencies may be used.

Alternatively an HF receiver may be used on the higher frequency bands by means of a convertor. This is a separate unit consisting of an RF stage, a mixer and crystal-controlled local oscillator. Typically this may convert 144–146MHz to 28–30MHz, the output from the mixer at this frequency being taken to the antenna input of a receiver which tunes over 28–30MHz.

Such convertors are commercially available to convert the popular VHF and UHF bands to a number of different output frequencies.

Direct-conversion receivers

The basis of the superhet is the conversion of all signals to a fixed intermediate frequency, followed by demodulation and AF amplification to drive a loudspeaker. The incoming signals can, however, be converted directly to AF, and thus a much simplified type of superhet results. This is known as the 'synchrodyne'. It is not a new principle but has become popular in amateur radio as the direct-conversion receiver.

The local oscillator operates very close to the signal frequency so that the output of the mixer (which is equivalent to a product detector) is in the AF range. This is normally followed by a low-pass filter to restrict the audio bandwidth to about 3kHz and a high-gain audio amplifier to drive a loudspeaker.

The mixer is usually preceded by a simple untuned RF stage. Due to the difficulties of making a sufficiently stable

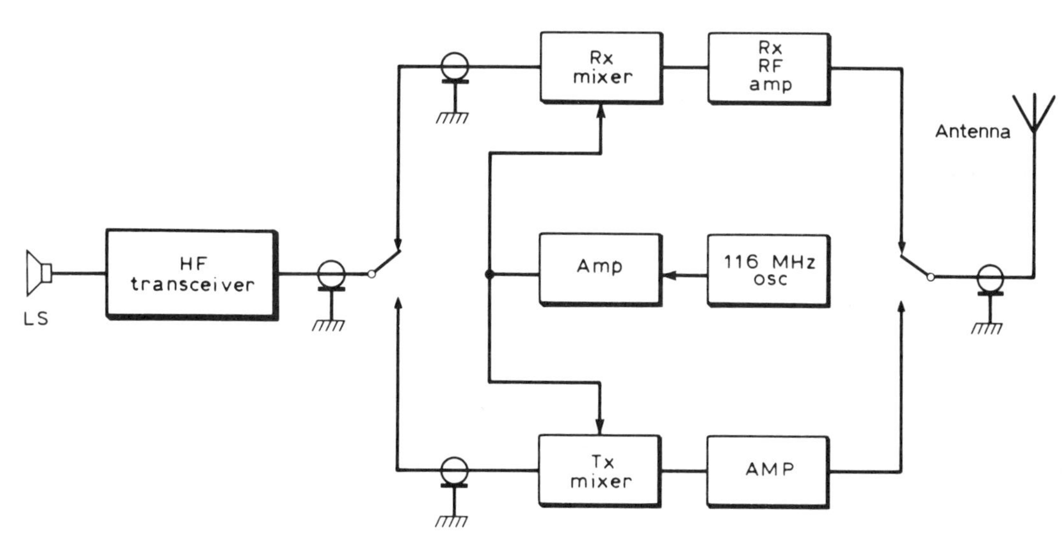

Fig 5.14. Block diagram of 28MHz/144MHz transverter

oscillator, the direct-conversion receiver is often restricted to the lower-frequency amateur bands where it is capable of surprisingly good performance, although obviously not in the same class as a good-quality communication receiver.

The oscillator followed by a buffer amplifier can also be used to drive a power amplifier, thus creating a low-power transceiver, and a number of these are now available commercially.

CHAPTER 6

Power supplies

Direct supplies of up to 12V which are required by solid-state circuits may be obtained from batteries of primary or secondary cells (for example, in mobile equipment). Fixed equipment is most conveniently served by a power unit which transforms, rectifies and smooths the public 240V 50Hz supply.

Rectifying circuits

Silicon diodes are used as the rectifying elements. Fig 6.1 shows the half-wave rectifying circuit in which current flows in the transformer secondary circuit (ie load, rectifier and secondary winding) only during the positive half-cycle. If the diode is reversed, then it is the negative half-cycle which is rectified.

Fig 6.2 shows the full-wave rectifier circuit. The diodes conduct on alternate half-cycles, and so this is a combination of two half-wave circuits. The load current waveform varies considerably in amplitude but, as it does not change polarity, it is a direct current.

This variation in the amplitude of the output voltage (or current) of a rectifier circuit is known as the 'ripple'.

The frequency of the ripple in the full-wave arrangement is 100Hz, while in the half-wave circuit it is 50Hz.

Fig 6.3 shows the bridge rectifier circuit. The output is full-wave, ie 100Hz ripple. At any one instant two of the diodes are in series carrying current. The transformer secondary winding does not have a centre tap and is required to supply a voltage of V_{ac} only (compare the full-wave arrangement of Fig 6.2 where the transformer supplies $2 \times V_{ac}$).

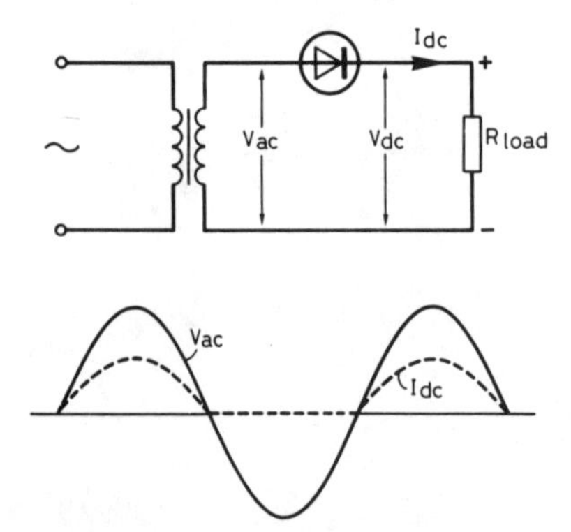

Fig 6.1. Half-wave rectifier and waveforms

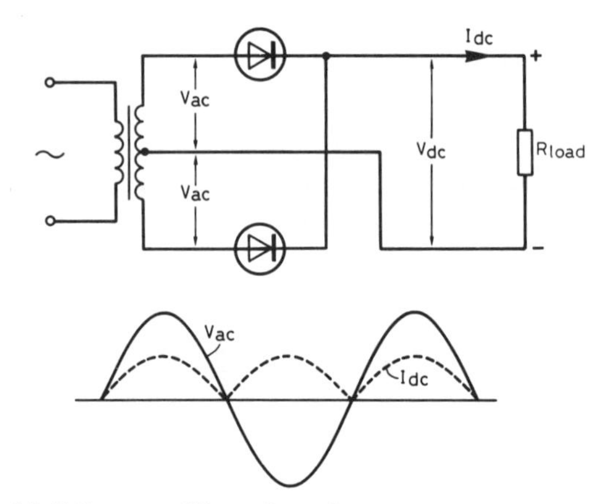

Fig 6.2. Full-wave rectifier and waveforms

Reservoir capacitor

C in Fig 6.4 is called a 'reservoir capacitor'. Its purpose is to store energy during the positive half-cycle and to supply the load during the negative half-cycle while the diode is non-conducting. The diode only conducts during the time V_{ac} exceeds the voltage across the reservoir capacitor (see Fig 6.5). If the value of C is made large, say 10,000μF, then a large pulse of current will be needed to charge it, because the time during which the diode conducts is short.

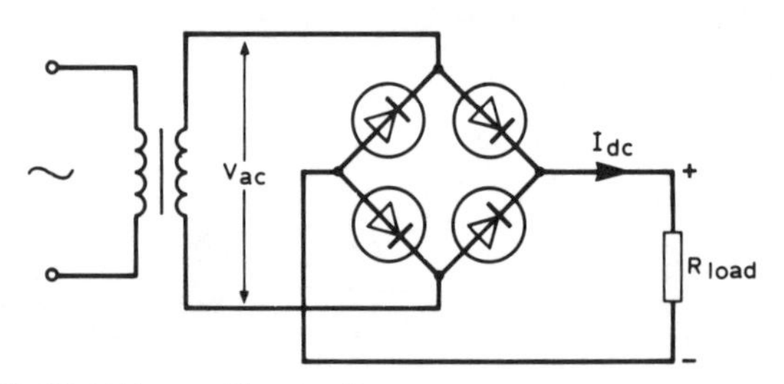

Fig 6.3. Bridge rectifier circuit

The reservoir capacitor charges up to a voltage which equals the peak value of V_{ac}, ie $\sqrt{2}$ times V_{ac}. On the negative half cycle, polarities are reversed but the magnitude is the same. Thus the maximum voltage across the diode is now 2 $\times \sqrt{2}V_{ac}$. This value is known as the 'peak inverse voltage' (PIV). The maximum PIV allowable across a rectifier diode is a very important characteristic of the diode.

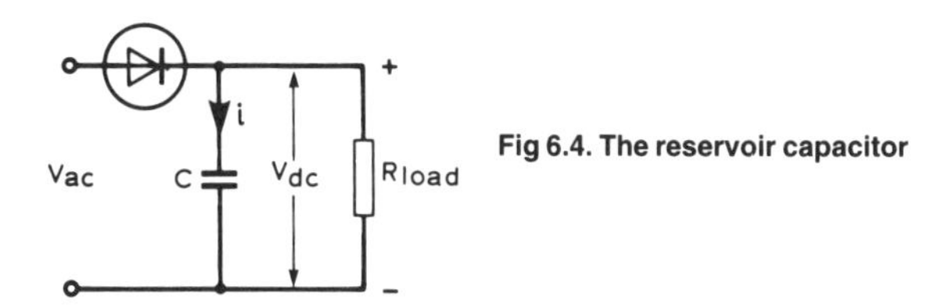

Fig 6.4. The reservoir capacitor

Smoothing circuits

By adding a choke L_s and a capacitor C_s as in Fig 6.6, the fluctuations in V_{dc}, ie the ripple, can be greatly reduced or 'smoothed'. In fact the output of the rectifier circuit as shown in Fig 6.5 consists of a direct voltage with an alternating voltage, ie the ripple, superimposed upon it. Thus L_s functions as a smoothing choke by opposing the alternating voltage and C_s provides a low-impedance path to earth for this voltage.

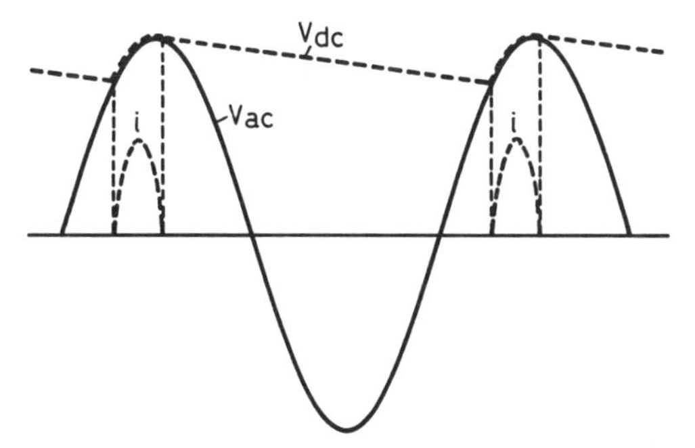

Fig 6.5. Output waveforms of rectifier circuit using a reservoir capacitor

A low-value resistor can be used in place of the choke but is not so effective. In the high-current supplies often required by transistor circuits, a smoothing choke is unacceptably large and expensive. In this situation a single high-value capacitor (eg 68,000μF) would be used for smoothing.

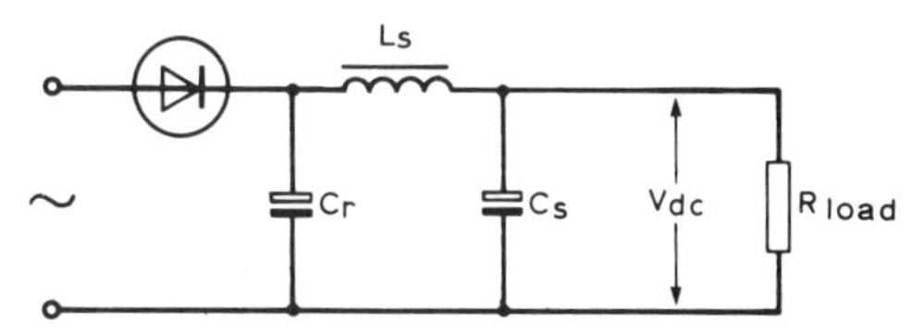

Fig 6.6. Smoothing circuit

A power supply may contain other components as in Fig 6.7. The primary circuit contains the mains switch and a fuse. A metallic screen, placed between the primary and secondary windings, is connected to the earth terminal. This screen helps to reduce mains-borne interference, and to protect the secondary winding from the voltage on the primary, should a short-circuit occur between them. This screen is generally made of thin copper (0.1mm thick) and of course is not continuous, otherwise it would act as a short-circuited winding. C1 may have a value of 1000μF and C2 4700μF. R1 will help with smoothing but must have a comparatively low value to avoid excessive voltage drop across it. The light-emitting diode indicates the ON condition. The resistor R2 will be 470Ω ¼W. Note the symbol used in Fig 6.7 for a bridge rectifier.

Properties of silicon diodes

Silicon diodes have a low internal voltage drop and can rectify large currents. A rapid increase in voltage, eg a switching transient, may give rise to a large current pulse, especially if the reservoir capacitor has a large value, and such a 'surge' is likely to destroy a silicon diode.

The capacitor C (about 0.01μF) in Fig 6.8 reduces the amplitude of the 'switching' transient (a short, but often high surge of current) when the diode turns off rapidly under bias conditions. The resistor R reduces the amplitude of the surge of current through the diode as the power supply is switched on and each time the reservoir (smoothing) capacitor charges up. The value of R must be chosen carefully to avoid an excessive voltage drop across it.

The PIV on each diode in a rectifier circuit depends on the circuit configuration and to a certain extent on the smoothing arrangement. In the half- and full-wave circuit, the PIV may be taken as approximately three times the direct output voltage of the lower supply and in the bridge circuit it is one and a half times the output voltage. The mean current in each

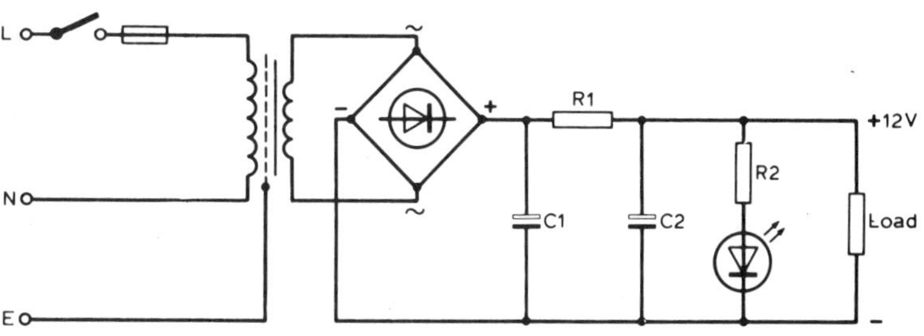

Fig 6.7. Practical power supply circuit

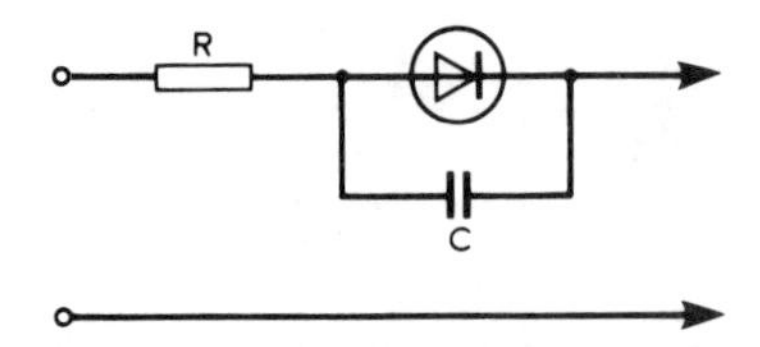

Fig 6.8. Surge protection diode

diode in the full-wave and bridge circuits is equal to one half of the DC output current.

Silicon diodes are commonly available with PIV from 50V up to about 1000V. Mean current capability varies from several amperes at the low PIV to an ampere or more at the higher voltage. Several diodes may be connected in series to provide a higher PIV capability. If this is done, a resistor (330kΩ) would be connected in parallel with each diode to equalise the voltage drop across each diode under reverse bias conditions.

Diodes and diode chains used in rectifying circuits should be very conservatively rated, eg five 50V diodes in series should be taken as a 200V diode.

Power supply characteristics

Most electronic circuits require a power supply which is very smooth, ie there is very little ripple voltage superimposed on the direct output voltage. Many circuits (particularly oscillators) require a supply which has an almost constant voltage irrespective of the amount of current being taken from the supply. Such a supply is said to have good 'regulation'. Variation of the mains voltage will also cause the output voltage to change.

Variation of the output voltage of a power supply is caused by the fact that the load current also flows through the power supply circuits which inevitably have a certain resistance or impedance. Thus the ideal power supply has a low 'source impedance' (cf the internal resistance of a cell). Source impedance can be minimised by careful design, but a more effective solution is the use of a regulated supply which also compensates for input voltage changes.

Stabilised power supplies

The simplest voltage regulator or stabiliser uses a zener diode as described in Chapter 3 (Fig 3.17). Reasonably constant supplies of up to 150V can be obtained in this way, and this

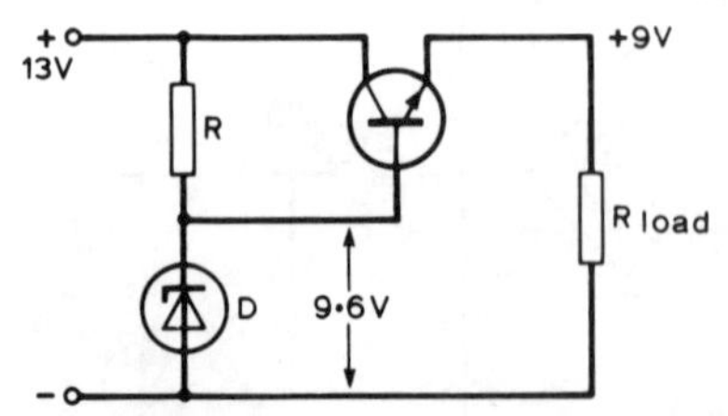

Fig 6.9. Simple voltage regulator

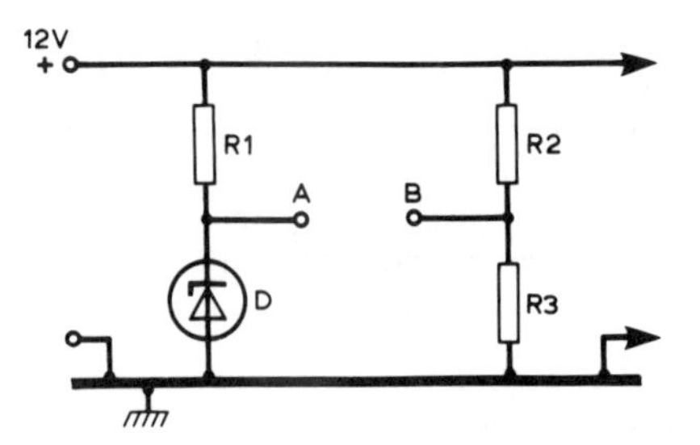

Fig 6.10. Voltage regulator – basic balancing circuit

circuit is often used for the supply of a VFO. A more effective method is to use a transistor as a regulator (the series or pass transistor) in series with the supply to the load as shown in Fig 6.9.

The base of the transistor is kept at a fairly constant forward voltage by the zener diode D. If the load increases, the voltage at the emitter will tend to fall. This in turn allows the transistor to conduct more easily and thus maintain the output voltage. In order to maintain better control a balancing (bridge) circuit is used, with an extra transistor working as a DC amplifier. Let the voltage in Fig 6.10 divide equally between R2 and R3. The voltage between point B and chassis will be 6V. If the diode D is rated at 6V then there will be no difference of potential between points A and B. If the line voltage falls then the voltage across R3 will fall, but the voltage across D will remain at 6V. The voltage at A will now be higher than the voltage at B. This change of voltage can be applied to the base of a transistor and so provide a degree of control. Fig 6.11 shows a circuit using such a device.

If the voltage at C increases the voltage at point B will increase. The emitter voltage of TR2 stays constant due to the action of D. Due to the increased base-emitter voltage in TR2, the collector current (and hence the voltage drop in R) will increase. The base-emitter voltage of TR1 is reduced, thus increasing the effective series resistance of TR1. This brings down the emitter voltage of TR1 and so point C returns to its normal voltage.

The small capacitors C1 and C2, about 0.1μF, prevent

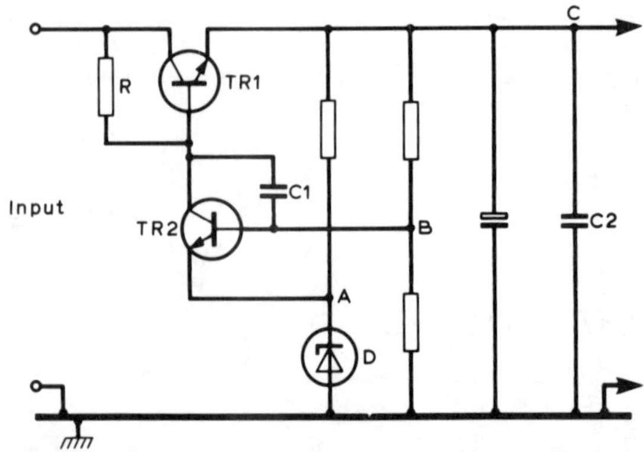

Fig 6.11. Voltage regulator using balancing circuit

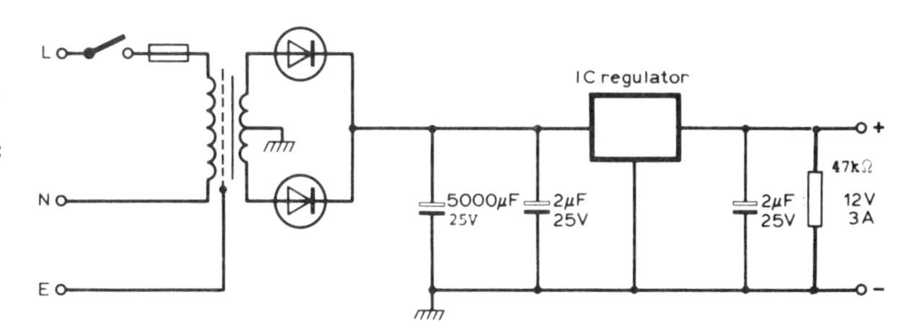

Fig 6.12. Practical power supply unit using IC regulator

instability (self-oscillation) of the output voltage if the load is constantly varying. There are many refinements in regulated supplies and manufacturers' literature gives much additional information.

The integrated-circuit stabiliser

The modern tendency in stabilised power supply design is to use an integrated circuit (IC) regulator. This contains in one small package the series element, reference voltage supply, a high-gain error amplifier and various sensing resistors and transistors. The more complex versions also contain protection circuits against too much voltage being applied to them and excess temperature rise. They are relatively cheap and are available up to a maximum regulated current of 10A. They may be connected in parallel for higher currents.

The simplicity of this type of power supply may be seen from Fig 6.12. The layout of such a power supply can be critical and, in particular, the connections between the integrated circuit and the 2μF capacitors should be short to avoid any instability arising from parasitic oscillation.

Safety note

Safety precautions are important in the use and servicing of any power supply. It must be remembered that low voltage power supplies operate from a 240V supply and that smoothing capacitors in such supplies do store a large amount of electrical energy.

Power supplies of the voltage discussed below should be regarded as lethal devices and treated accordingly.

See Chapter 12 for recommendations on the safety precautions to be observed when dealing with power supplies.

Power supplies for RF linear amplifiers using valves

The waveform of the SSB signal has a low average value and a high peak value as it represents the syllabic variations of the speech waveform. Thus the standing (no signal) current drawn by the SSB linear amplifier may be of the order of 30–40mA, and this is likely to rise to a peak value of around 250mA.

The voltage required will be about 600–700V in the case of an amplifier using TV line output valves and 1500–2300V if valves such as the 572B or 813 are used. This supply voltage should not fall by more than 5% under the peak load referred to above.

The circuit generally used is the bridge rectifier arrangment shown in Fig 6.3. According to the PIV of the diodes used, several in series may be needed in each arm of the bridge. Across each diode should be a 0.01μF capacitor for surge protection and a resistor (order of 330kΩ) to equalise the voltage across the each diode.

Smoothing is by means of a very large value reservoir capacitor, in fact it is this capacitor which supplies the peak load. This capacitor would be made up of a bank of series or series-parallel electrolytic capacitors according to availability with a total capacitance of not less than about 40μF. A resistor (330kΩ) across each capacitor equalises the voltage across each one and also serves as an HT bleed to discharge the capacitors on switch-off.

In order to maximise the reliability of such a power unit, the rating of each resistor, capacitor and diode should be such that it is under-run by a factor not less than 25%.

Power supplies of the voltage discussed above should be regarded as lethal devices and treated accordingly.

CHAPTER 7

Propagation and antennas

Radio communication depends on the radiation of electromagnetic waves from the transmitting antenna. The electromagnetic waves are created by the alternating RF currents in the antenna which arise from the coupling of the output of the transmitter into the antenna system.

The transmitted signal may be regarded as a succession of concentric spheres of ever-increasing radius, each one a unit of one wavelength apart, formed by forces moving outwards from the antenna. These hypothetical spherical surfaces, called 'wave-fronts', approximate to plane surfaces at great distances.

There are two inseparable fields associated with the transmitted signal, an 'electric field' (E) due to voltage changes and a 'magnetic field' (H) due to current changes, and these always remain at right-angles to one another and to the direction of propagation as the wave proceeds. The oscillations of each field are in phase and the ratio of their amplitudes remains constant. The lines of force in the electric field run in the plane of the transmitting antenna in the same way as would longitude lines on a globe having the antenna along its axis. The electric field is measured by the change of potential per unit distance, and this value is termed the 'field strength'.

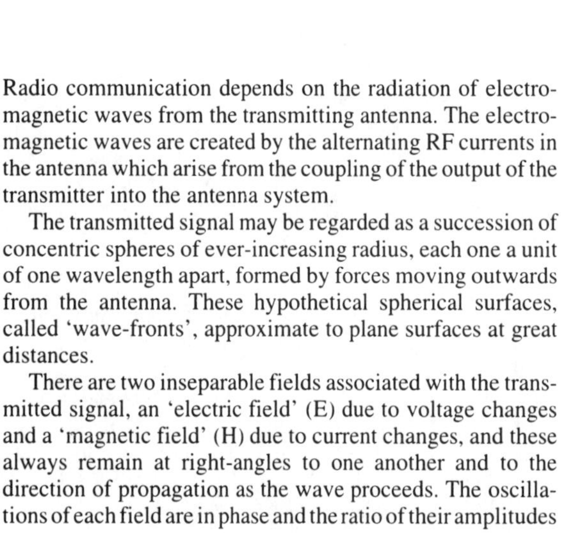

The two fields are constantly changing in magnitude and reverse in direction with every half-cycle of the transmitted carrier. As shown in Fig 7.1, successive wave-fronts passing a suitably-placed second antenna induce in it a received signal which follows all the changes carried by the field and therefore reproduces the character of the transmitted signal. The field strength at the receiving antenna may range from less than 1μV/m to greater than 100mV/m.

Waves are said to be 'polarised' in the direction of (parallel to) the electric lines of force. Normally the polarisation is parallel to the length of the antenna, ie a horizontal antenna produces horizontally polarised waves. In order to receive maximum signal strength, the receiving antenna must be orientated to the same polarisation. In practice, particularly at VHF, the polarisation may be modified by factors such as abnormal weather conditions and reflection from the ionosphere.

The electromagnetic wave is an alternating quantity. Its wavelength (λ) is the distance, in the direction of propagation, between points where the intensity of the field is similar in magnitude and sign, ie the distance travelled in space to complete one cycle. Therefore:

$$\text{velocity} = \text{frequency} \times \text{wavelength}$$
$$c = f \times \lambda$$

where c is the velocity of propagation which for electromagnetic waves in space is approximately 300,000,000m/s (186,000 miles/s). Therefore:

$$\lambda(\text{m}) = \frac{300{,}000{,}000}{f(\text{Hz})} = \frac{300}{f(\text{MHz})}$$

Fig 7.1. The fields radiated from a transmitting antenna. (a) The expanding spherical wavefront consists of alternate reversals of electric field, with which are associated simultaneous reversals of the magnetic field at right angles to it, as shown in (b) and (c). The dotted arcs represent nulls. The lower diagrams should be interpreted as though they have been rotated through 90° of arc, so that the magnetic field lines are perpendicular to the page

Modes of propagation

The three main modes of propagation of electromagnetic waves are:

(a) ground (or surface) wave;
(b) ionospheric wave (sky wave);
(c) tropospheric wave.

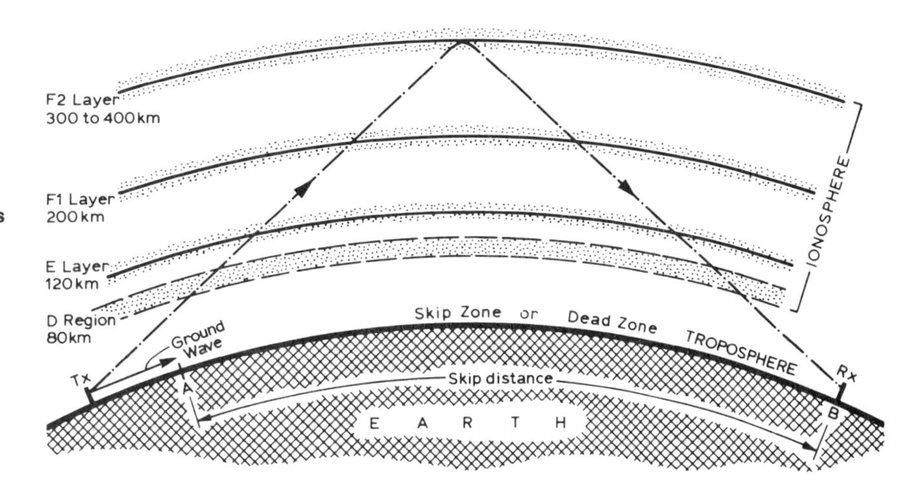

Fig 7.2. Reflection of radio waves by ionised layers

Ground-wave propagation

In ground-wave propagation, the radiated wave follows the surface of the earth. It is the major mode of propagation for frequencies up to 1MHz to 2MHz. Attenuation of the ground wave increases very rapidly above 2MHz and it may extend for only a few kilometres at frequencies of the order of 15–20MHz. At very low frequencies the attenuation decreases to such an extent that reliable world-wide communication is possible at all times. The ground wave is not so affected by atmospheric effects or time of day as other modes, particularly at frequencies below about 500kHz.

Ionospheric propagation

Ionospheric propagation is the 'refraction' (ie bending), and hence reflection, of radio waves back to earth by layers of ionised gases as shown in Fig 7.2. It is the normal mode of propagation over the frequency range of about 1MHz to 30MHz.

These layers are the F2 layer (height 300–400km); F1 layer (about 200km) and the E layer (about 120km). At night and in midwinter, the F1 and F2 layers tend to combine into a single layer at a height of about 250km. At about 80km there is a much less distinct layer which is generally known as the D region.

The ionised layers are the result of the ionisation of the oxygen, nitrogen and nitric oxide in the rarefied atmosphere at these heights by X- and ultra-violet radiation of various wavelengths which comes from the sun.

When these gases are ionised the molecules split up into ions and free electrons, and recombine after sunset. This whole region is therefore known as the 'ionosphere'.

The solar radiation which causes the ionisation is continually varying; hence the degree of ionisation varies considerably according to season and time of day. It has also been found that the degree of ionisation is affected by the number of sunspots.

The number of sunspots varies cyclically, with maximum activity occurring at about 11-year intervals. Thus maximum ionisation occurs at the same intervals. The next maximum may not occur until 2002-4. As the frequency of the radio wave increases, a greater level of ionisation is needed to cause reflection. The F2 layer normally has the greatest ionisation and so it is the F2 layer which reflects the highest frequencies which have passed through the lower layers. It is seen from Fig 7.2 that it is this layer which reflects back to earth at the greatest distance from the transmitter. Therefore it is the characteristics of the F2 layer which are of most interest and significance in long-distance communication. The major significance of the D region is that it absorbs the frequencies under discussion in abnormal circumstances.

The maximum frequency which is reflected in the ionosphere is known as the 'maximum usable frequency' (MUF). This frequency depends on many factors, ie season, time of day, path latitude and state of the sunspot cycle. Signals above the MUF pass through the F2 layer and are lost in space. The curves of Figs 7.3–7.6 indicate the likely variation of the MUF, as follows.

1. The peak value of the MUF generally occurs between 1000 and 1600 hours.
2. Peak values are much higher at sunspot maximum than at the sunspot minimum.
3. Peak values are much higher in the winter than in the summer.
4. There is a much larger variation in the MUF over the day in the winter than in the summer.
5. Comparison of Figs 7.3 and 7.5 shows that the MUF is higher in a north-south direction, eg the London-Cape Town path than in an east-west direction, eg London-New York path. Figs 7.4 and 7.6 show MUF variations for intermediate directions.

Around the sunspot maximum, the MUF may exceed 50MHz for short periods, but at the minimum it rarely exceeds 25MHz.

Mainly during the summer months, regions of intense ionisation may occur in the E layer which is therefore able to

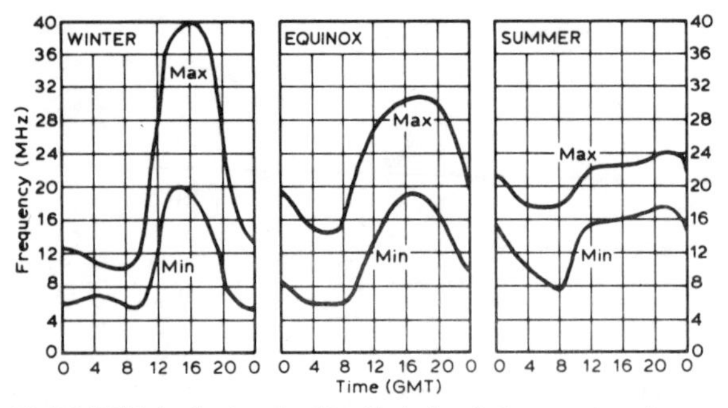

Fig 7.3. MUFs for the London-New York circuit at sunspot maximum and minimum

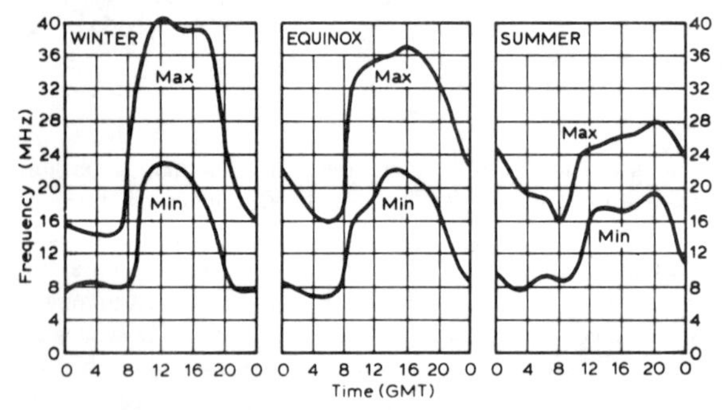

Fig 7.4. MUFs for the London-Buenos Aires circuit at sunspot maximum and minimum

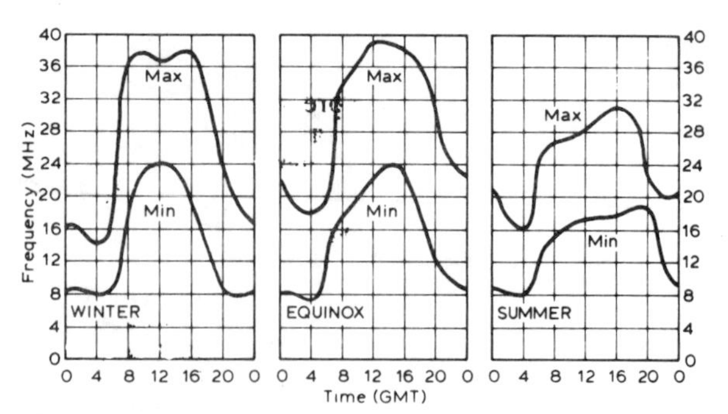

Fig 7.5. MUFs for the London-Cape Town circuit at sunspot maximum and minimum

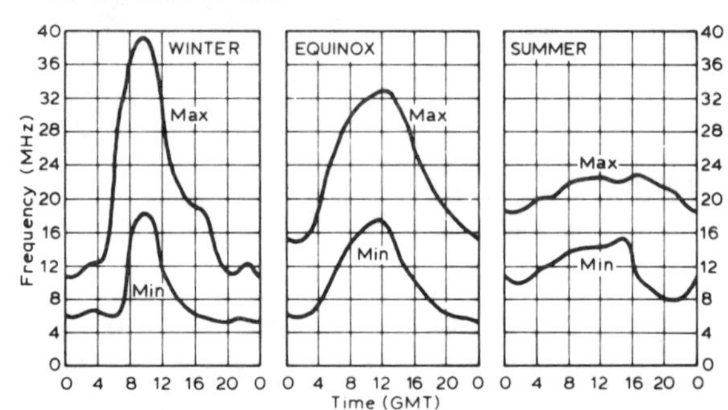

Fig 7.6. MUFs for the London-Chungking circuit at sunspot maximum and minimum

reflect much higher frequencies than normal, ie up to 100MHz and occasionally 150MHz. This ionospheric propagation can occur in the 50, 70 and 144MHz bands. This is known as 'sporadic E' propagation; it often also causes extremely strong signals with deep fading, particularly on 28MHz.

The 'critical frequency' is the highest frequency reflected when the radiation is vertical. This frequency is lower than the MUF and will be different for each layer. The forecasting of MUF from daily measurements of critical frequency made at radio observatories all over the world is of great importance in commercial communications. Forecasts are made for several years ahead and are continually refined as later measurements become available.

There is no simple explanation of the many anomalies in the behaviour of the F2 layer and most of what is known is based on experimental results and deduction. As far as amateur radio is concerned, it is convenient to accept the published variations of MUF in particular as of most significance to communication on the amateur bands. The fact that the MUF is highest in the early winter months should be noted.

It is clear from Fig 7.2 that there is a region between the transmitter and the point at which the reflected wave returns to earth (B) where no signal is received. This is the 'skip zone' or 'dead zone'. However, there will be inevitably some ground-wave propagation associated with the transmission and hence, more accurately, the 'skip distance (zone)' starts where the ground wave has decayed to zero, ie (A) in Fig 7.2.

The maximum distance along the surface of the earth which results from a single reflection from the F2 layer is about 4000km (2500 miles); thus world-wide communication implies several reflections from the F2 layer to earth, back to the F2 layer, and so on.

Communication by ionospheric propagation may be disturbed or interrupted by abnormal radiations from the sun, especially in the period soon after a sunspot maximum. Intense solar flares (ie eruptions at the surface of the sun) greatly increase the ultra-violet and X-radiation from the sun. This has the effect of greatly increasing the level of ionisation in the D region and results in the absorption of radio waves before they reach the reflecting layers, and thus there can be a complete interruption of communication ('Dellinger fade-out') over all or part of the HF spectrum which may last for a few minutes to an hour or so. This is known as a 'sudden ionospheric disturbance' (SID).

An SID may be followed about two days later by another form of fade-out or blackout, the 'ionospheric storm', and this can last from a few hours to several days. It is thought that ionospheric storms are caused by slower-moving particles, emitted at the same time as the solar flare, which cause increased ionisation in the D region but decreased ionisation in the F layer.

Fading of a signal propagated ionospherically, as opposed to the fade-out described earlier, is a common occurrence. The signal received at a given point is rarely constant because of the continually changing conditions in the ionosphere, ie layer height, ionisation level and possibly skip distance if the frequency is close to the MUF.

It is also possible that the signal may arrive by two different paths, ie by one reflection and also by two reflections; in this case, the time delay between the different paths may cause distortion.

The effects of fading may be minimised by really effective automatic gain control in the receiver.

Tropospheric propagation

This is the major mode of propagation over long distances (ie beyond the line-of-sight range) at frequencies above about 50MHz.

The troposphere is the name given to the lower part of the atmosphere. Its height varies from about 6km to about 17km and depends upon latitude and atmospheric pressure. Changes in temperature, pressure and humidity of the atmosphere (ie weather changes) cause large changes in its refractive index at increasing height above the earth's surface (refractive index is a quantity which is a measure of how much a radio wave is bent as it passes through the atmosphere).

These changes in the refractive index affect the propagation within the troposphere of waves of frequencies above approximately 40–50MHz in a number of ways, as follows.

1. Localised variations cause scattering of radio waves.
2. Sharp changes between horizontal layers cause reflection (cf ionospheric propagation).
3. A sharp decrease in refractive index with height can create the phenomenon of 'ducting'.

A duct is a region of indeterminate shape which may cover a very large area but only be 40–50m high. It has the property of propagating radio waves with extremely low attenuation, and such waves therefore tend to hug the earth's surface. A duct may last for several days.

A wave which gets 'trapped' in such a tropospheric duct can travel for very long distances (1500km or more) but can leak out at any point.

This is not a reliable mode of propagation; it can cause severe interference to very distant services. However, it is of very great interest in amateur radio as it enables long-distance contacts on the VHF and UHF bands to be made with very low power and simple antennas.

Periods of enhanced tropospheric propagation can often be forecast by observation of weather changes.

The mode of propagation depends on the frequency used, but there is no sharp transition from one mode to another as the frequency increases. This depends on many factors and at some frequencies significant propagation can occur by more than one mode. For example, long-distance propagation on the medium-wave broadcast band and the 1.8MHz amateur band during the hours of darkness is by sky wave.

Antennas

The fundamental antenna is a piece of wire which is one half of a wavelength (λ/2) long, corresponding to the frequency at

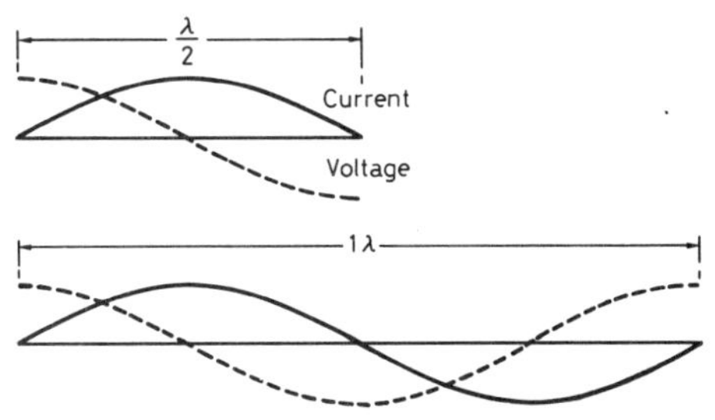

Fig 7.7. Standing waves on resonant antenna showing voltage and current variation. Upper antenna is λ/2 long (fundamental antenna); lower is 1λ (second harmonic) antenna

which radiation is desired. The voltage and current vary over the length of this antenna, as shown in Fig 7.7 (top).

If the piece of wire is made a whole wavelength (1λ) long, the current and voltage variations are as in Fig 7.7 (bottom). This is known as a 'full-wave' or 'second-harmonic' antenna. Larger multiples of the basic λ/2 antenna show similar voltage/current variations. These variations are known as 'standing waves', and this type of antenna is known as a 'resonant' antenna.

It is seen that the ratio of voltage and current varies over the length of the antenna, and may be resistive, inductive or capacitive. This ratio is referred to in general terms as the antenna 'impedance'.

The 'radiation resistance' of an antenna is a fictitious resistance which would dissipate the power radiated by it.

Antenna length

The length of a half-wavelength (λ/2) in space is

$$\frac{150}{f(\text{MHz})} \text{ metres}$$

The actual length of a λ/2 antenna is somewhat less than this, owing to:

(a) the velocity of propagation in the wire being different from that in space;
(b) the presence of insulators at the end of the wire and of nearby objects (trees or buildings);
(c) the diameter of the wire or element.

Table 7.1. Approximate lengths of λ/2 dipoles

Band (MHz)	λ/2 (m)	λ/2 (ft)
1.8	75.2	247
3.5	39.2	129
7	20.3	67
10	14.1	46
14	10.05	33
18	7.9	26
21	6.7	22
24	5.7	18.8
28	4.93	16.2
50	2.8	9.3
70	2.02	6.6
144	0.97	38.4in
430	0.32	12.8in

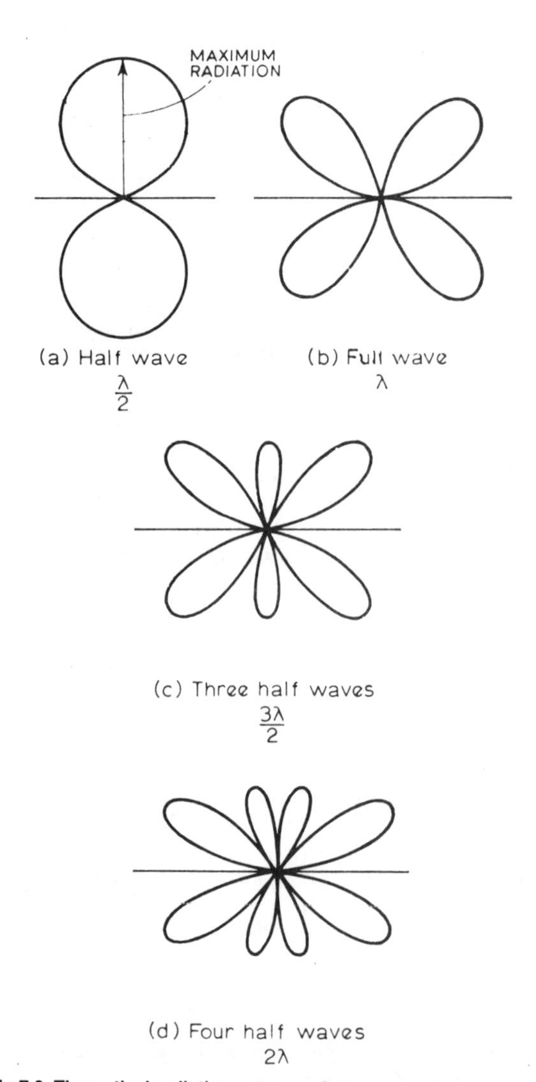

Fig 7.8. Theoretical radiation patterns of resonant antennas

The actual length is normally taken to be 5% less than (or 0.95 of) the electrical length. This constant, 0.95, is sometimes known as the 'correction factor', hence the actual length is

$$\frac{150 \times 0.95}{f(\text{MHz})} = \frac{143}{f(\text{MHz})} \text{ metres}$$

Radiation patterns

If a λ/2 antenna is assumed to be parallel to and at least a wavelength above perfect ground, and also remote from all other objects, the radiation is concentrated at right-angles to its length, as shown in Fig 7.8(a). This is the radiation pattern of a λ/2 antenna and, as the antenna radiates in directions all round the wire, the radiation pattern in space is the shape formed by imagining the pattern of Fig 7.8(a) to be rotated round the antenna as an axis.

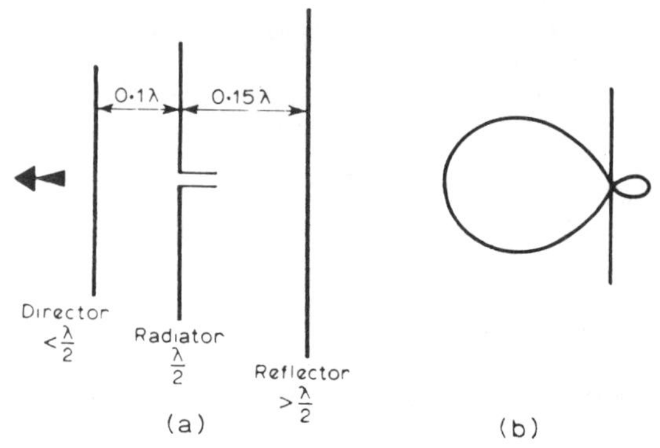

Fig 7.9. (a) Arrangement of Yagi directional (beam) antenna. (b) Radiation pattern of directional antenna

Radiation patterns of antennas working on higher harmonics are shown in Figs 7.8(b), (c) and (d), where it is seen that the effect is to produce more lobes; the four lobes of the full-wave (1λ) case tend to swing towards the ends of the antenna, and subsidiary lobes appear. Thus in the case of an extremely long antenna the radiation tends to be concentrated towards the ends.

If one end of the antenna is tilted, the lobes tend to move together and hence radiation tends to become concentrated off the lower end. Consideration of the λ/2 pattern shows that radiation is all around (omnidirectional) when the antenna is vertical.

Angle of radiation

This is the angle with respect to a tangent at the surface of the earth at which the maximum radiation occurs. Its value depends upon interaction between the direct ray from the antenna and the ray reflected in particular from the ground. Hence it is determined by both the antenna height and the characteristics of the ground. A single lobe exists in the vertical plane at an antenna height of approximately λ/2, and above this height the one lobe splits into two, one at a higher angle and the other at a lower angle. These two lobes could be felt to be of more use than a single lobe from the point of view of total coverage (see Fig 7.2). As the height increases the higher lobe increases in angle and magnitude. It is clear that the angle of radiation produced by a multi-band antenna at the fairly average height of 7–8m varies considerably between 3.5MHz and 28MHz.

Directional antennas

It is possible to modify the radiation pattern of an antenna in order to concentrate the radiation in a particular direction. Thus a 'directional' or a 'beam' antenna is created.

This is generally achieved by the addition of parasitic elements known as a 'director' and a 'reflector' parallel to the antenna, as shown in Fig 7.9(a), which also shows the approximate spacing in terms of the operating wavelength. This arrangement is known as the 'Yagi array'. The radiation pattern is shown in Fig 7.9(b). The addition of more directors produces a narrower beam, but more than one director is usually only possible at VHF, where up to 20 or more directors may be used.

Means must be provided for rotating a beam antenna so that it can be turned to the required direction.

An additional feature of the directional antenna is the fact that when used for reception, signals to the back of the beam are attenuated, ie interfering signals from an unwanted direction may be significantly reduced in strength. The characteristics of a beam antenna are the 'forward gain' (compared with a dipole) and 'front-to-back ratio', these terms being self-explanatory; they are expressed in decibels.

Feeders and transmission lines

For obvious reasons, antennas should be as high and as far away from buildings etc as possible. It may therefore be necessary to transfer power from the transmitter to the antenna over a fairly long distance, say 10 to 25m. How this is done depends on the type of antenna used. The general name for this connection is 'feeder', which can be just a single wire or a transmission line which may be 'balanced' or 'unbalanced', both types consisting of two conductors. In a balanced line, both conductors have equal potential to earth, ie neither is earthed. In the unbalanced line, one conductor is earthed.

Connections of this type are subdivided according to a property known as the 'characteristic impedance' (Z_0) which is measured in ohms. The characteristic impedance of a balanced line depends upon the diameter of the conductors and the spacing between them. Balanced or 'twin feeder' is commercially available with impedances of 75 and 300Ω (300Ω ribbon). Open-wire feeders of 300 to 600Ω impedance can be made by spacing apart two lengths of wire of 14swg (2mm) or 16swg (1.6mm) by low-loss spacers tied to the wire at intervals of 30–40mm. The unbalanced line is the familiar coaxial cable, where the characteristic impedance now depends upon the diameter of the conductor (the 'inner') and the internal diameter of the screen (the 'outer') which is earthed. The common impedances are 50 and 75Ω.

The velocity of propagation of an electromagnetic wave in a transmission line is less than in free space. The ratio of the velocities is the 'velocity ratio' or 'velocity factor'. For most solid polythene-insulated coaxial cables the velocity ratio is about 0.66; 300Ω twin feeder has a velocity ratio of about 0.85.

For the optimum transmission of power, a transmission line must be 'matched', ie 'terminated', at the load end by the correct impedance; this is equal to the characteristic impedance of the line.

Standing waves

In a transmission line which is not correctly terminated, ie the load impedance is not equal to its characteristic impedance, an input wave (the 'incident' or 'forward' wave) is reflected back to the input end. This is the 'reflected' or 'reverse' wave which is smaller in amplitude than the forward wave. It cannot cancel out the forward wave but combines with it to create points of maximum and minimum voltage (and current).

These variations of voltage (and current) in a mismatched transmission line are known as 'standing waves'. The ratio of maximum and minimum voltage (or current) is the 'standing wave ratio'. Generally it is the voltage which is measured, leading to the term 'voltage standing wave ratio' (VSWR) which is usually abbreviated to 'SWR'. In a correctly terminated line, there is no reflection of the forward power at all, and the SWR is then 1 to 1.

Stubs

A line is completely mismatched when the far end is either a short-circuit or an open-circuit. There is then no load resistance to dissipate power, and 100% reflection of current and voltage occurs. It can be shown that a λ/4 length of a line which is open-circuit is equivalent to a series-tuned circuit at the frequency corresponding to the wavelength and therefore presents a very low resistance at that particular frequency. In a complementary fashion a short-circuit is equivalent to a parallel-tuned circuit and so presents a very high resistance.

These lengths of line are known as 'stubs'. The open-circuit stub is particularly useful as a means of creating a short-circuit at a particular frequency.

The velocity ratio must be taken into account when calculating the length of the line which is hence somewhat shorter then the free-space length.

Coupling the transmitter to the antenna

Modern commercial transmitters and transceivers have output circuits which are designed to 'look into' an unbalanced load of 50 to 70Ω. For optimum transfer of power to the antenna, impedances throughout the system must be matched. For example, if the antenna feed-point impedance is 50Ω, this should be connected to the transmitter output socket by an unbalanced line, ie coaxial cable of 50Ω impedance. The output power for which the transmitter is designed is then transferred to the antenna from which it is all radiated.

In practice the feed-point impedance of an antenna can vary widely from its nominal value. Two typical reasons are:

1. Siting conditions: the proximity of the antenna to local objects such as buildings and trees, and with HF antennas in particular, height above ground. The lower an antenna for a given band, the lower the impedance.
2. For a given antenna, excursions within a band, ie from one end to the other, will cause a change. Some types, in particular multi-band antennas, will vary more than others and also from band to band.

Hence, as the mismatch increases, the reflected (reverse) power rises and the SWR increases. Thus SWR is a measure of the effectiveness of the whole system.

Matching the feeder impedance to the antenna itself may not be a straightforward task, particularly in the case of multi-band antennas which are themselves a compromise.

The solution to this problem, which has become virtually

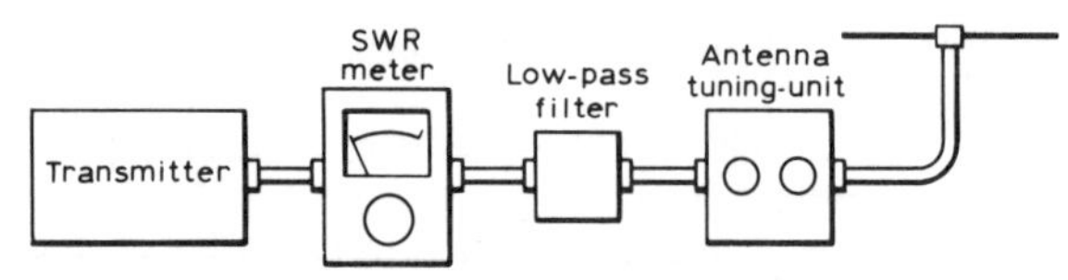

Fig 7.10. Preferred arrangement of transmitter-to-antenna circuit

standard practice, is to match the transmitter to the feeder plus antenna as shown in Fig 7.10 and, in more detail, in Fig 9.6 in Chapter 9, 'Electromagnetic compatibility'.

Matching is achieved by means of an antenna tuning unit (ATU). This does not tune the antenna; it matches it to the transmitter output and hence should really be called an 'antenna matching unit', but it has been known as an ATU by common usage for very many years. Other names now used are 'antenna system tuning unit', 'transmatch' or 'matching network'.

Thus the transmitter-to-ATU connection is a very short matched transmission line at 50/70Ω. This has the advantage that it can include means of measuring the SWR and a low-pass or other filter if necessary. These two items would be designed for use at the line impedance and so the filter would have its designed attenuation. The ATU is basically a tuned circuit and so provides about 25dB attenuation of harmonics and other unwanted frequencies in the transmitter output. It has an unbalanced input socket and generally it can provide both balanced and unbalanced outputs.

A basic circuit of an SWR meter is shown in Fig 10.7 in Chapter 10, 'Measurements'. This is the simple reflectometer type originated in the USA as the 'Monimatch'. The SWR meter and the ATU exist in many home-made and commercial versions.

This arrangement has become more significant in recent years due to the use of transistors in the transmitter output stage. Such transmitters often incorporate automatic means of switch-off or power reduction if the SWR rises above about 2.5 to 1. The more sophisticated transmitters now include an ATU and SWR meter in their output circuit.

This aspect is not so critical if the output stage uses valves. Earlier transmitters were capable of matching almost any length of wire to most amateur bands.

A perfectly matched system will have an SWR of 1 to 1. A modern commercial transmitter may switch off or reduce power at about 2.5 to 1. The question therefore arises: what is an acceptable maximum SWR? There are many possible errors in SWR measurement, and so an SWR of 1 to 1 could be regarded with a certain amount of suspicion. A system which appears to have an SWR greater than about 5 to 1 should certainly be investigated, although the power loss then is only just less than 3dB. It is probably more important to reduce SWR to safeguard a solid-state transmitter output stage than for any other reason.

In practice the consequences of SWR are:

1. Greater loss in the feeder. How much greater depends on feeder type and frequency. In general it can be said to be inconsequential at HF (up to 30MHz), but may be significant at VHF (144MHz) and will certainly be so at UHF (432MHz and upwards).
2. Use of very high power with an excessive SWR may cause breakdown of the feeder or units 'in line', such as filters or switches. Breakdown can be caused by flashover (due to high voltage) or melting of conductors or dielectric (due to high current).

Note that a high SWR, of itself, does not cause a feeder to radiate, or produce TVI or other interference.

The impedance matching of the transmitter/antenna system discussed above applies to the HF bands. At VHF and UHF, the transmitter and antenna are normally designed for single-band operation rather than for several bands. A much tighter control of impedance is therefore possible. Antenna length is much shorter compared with the height above ground, and hence there is less doubt about feed-point impedances. An SWR meter can be incorporated to check the integrity of the antenna system.

To summarise, current amateur technology is to match as closely as possible to the ATU input and to accept the SWR presented by the antenna to the ATU output. It may be difficult to measure this. It depends on how the antenna is fed: single wire, coaxial cable or open-wire feeder for example. The likely antenna performance can be judged fairly accurately by visual inspection, ie is it in the clear and at a reasonable height compared with the longest wavelength used? If it is, then it is likely to have a good performance, propagation conditions permitting! If the antenna is low and hemmed in by buildings and trees, it is then likely to be less good, although antennas so placed often perform surprisingly well.

Practical antennas

Antennas for use in the amateur bands are usually based on the fundamental antenna, ie the λ/2 dipole. Study of the usual textbooks on amateur radio reveals that many different forms of antenna have been developed. Basic information on the more common types only can be given within the scope of this manual.

The dipole

The impedance of a λ/2 antenna at the centre point is roughly 70Ω (resistive). Thus this point could be coupled directly to the output of a transmitter by 70Ω coaxial cable, resulting in a good impedance match (Fig 7.11). This arrangement is known as a 'half-wave dipole' or simply as a 'dipole'. Lengths of half-wave dipoles are given in Table 7.1.

As mentioned earlier, the centre impedance of the dipole depends upon the height of the antenna and the proximity of buildings etc. The arrangement of Fig 7.10 may therefore be preferable.

However, if a 7MHz dipole was fed with power at, say, 14MHz or 28MHz, there would be an impedance mismatch, as the impedance at the centre would no longer be 70Ω. It would be much higher than this and moreover would not be resistive.

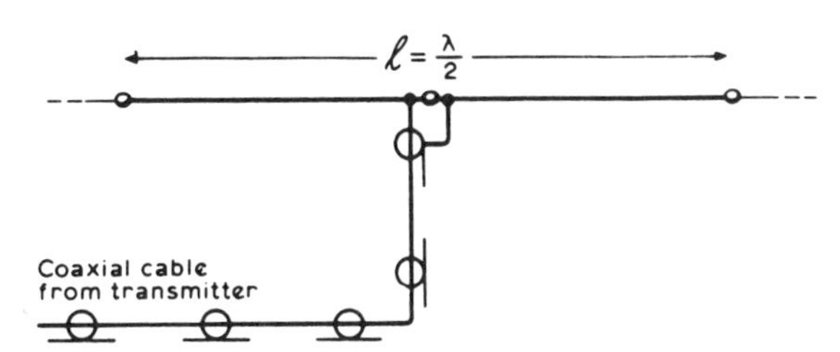

Fig 7.11. λ/2 dipole antenna fed by coaxial cable. Length is approximately equal to 143/*f* metres

The exception to this is that the impedance at the third harmonic is around 90Ω. The only application of this in amateur radio is the use of a 7MHz dipole at its third harmonic, ie in the 21MHz band. The mismatch is then not excessive.

The dipole is a satisfactory antenna but, apart from the example quoted above, it is a single-band antenna. The common use of coaxial cable (unbalanced) to feed a dipole is convenient as the output of most transmitters is unbalanced, but consideration of the antenna itself shows that a dipole is balanced so that it should not be fed with an unbalanced cable.

Alternative and more correct arrangements are either the use of 75Ω twin cable between the antenna and the ATU, or a balance-unbalance transformer between the top end of the coaxial feeder and the antenna.

The balance-unbalance transformer, commonly called a 'balun', enables a balanced circuit to be coupled to an unbalanced circuit and vice versa. In one version, it consists of three tightly coupled windings on a small ferrite core, as shown in Fig 7.12.

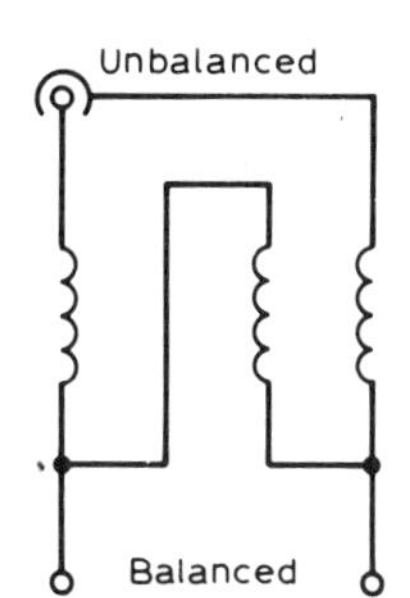

Fig 7.12. Arrangement of windings in balun transformer

The trap dipole

The trap dipole, sometimes known as the 'W3DZZ antenna' after the callsign of its originator, is a dipole having a parallel-tuned circuit or 'trap' inserted at a particular point in each leg as shown in Fig 7.13. At resonance, the trap presents high impedance and therefore at the resonant frequency the length beyond the trap is virtually isolated from the centre portion. Below the resonant frequency the trap provides an inductive reactance which reduces the length of antenna required for resonance.

Fig 7.13 gives the dimensions of the trap dipole; the traps are resonant at 7.1MHz. At 7MHz the system operates as a λ/2

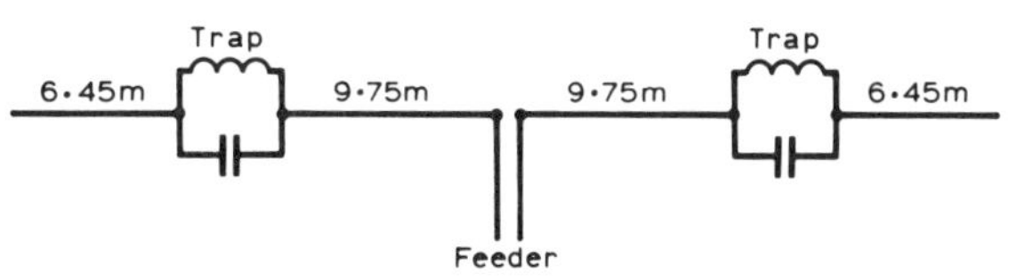

Fig 7.13. The trap dipole

dipole, the traps isolating the outer sections. At 3.5MHz it operates as a λ/2, the traps electrically lengthening the top. At frequencies above the resonance of the trap, the end sections are not isolated, but the traps do provide series capacitance. This enables the antenna top to resonate at odd harmonics of its fundamental and so at 14MHz, 21MHz and 28MHz the trap dipole functions as a 3λ/2, 5λ/2 and 7λ/2 antenna respectively. A reasonably satisfactory match to a 75Ω feeder is obtained on each band. At 1.8MHz the feeders may be joined together at the transmitter and the system will operate satisfactorily as a top-loaded Marconi antenna against ground or a counterpoise.

The trap dipole has become very popular as a multiband antenna in recent years because of the commercial availability of suitable traps. It must be appreciated that, as with all multi-band antennas, it is a compromise arrangement and as such will not give optimum results on every band.

The folded dipole

A dipole arranged as shown in Fig 7.14(a) is said to be 'folded'. This is the simplest example of folding and it results in the centre impedance being multiplied by four.

This increase in impedance is the main advantage of folding. A folded dipole has an impedance of about 300Ω and so may be fed with 300Ω twin feeder.

The loss in 300Ω feeder is somewhat lower than that in 75Ω coaxial cable and so a folded dipole may be preferred if the feeder length is extremely long. In fact, as shown in Fig 7.14(b), it is possible to construct a folded dipole entirely of 300Ω feeder.

The vertical antenna

A vertical antenna offers the attraction of low-angle, omnidirectional radiation and is popular where space does not allow a long horizontal antenna.

The simplest form is a vertical radiator one quarter of a wavelength (λ/4) long (see Fig 7.15); the impedance at the bottom is 30–40Ω and so it can be fed by 50Ω coaxial cable.

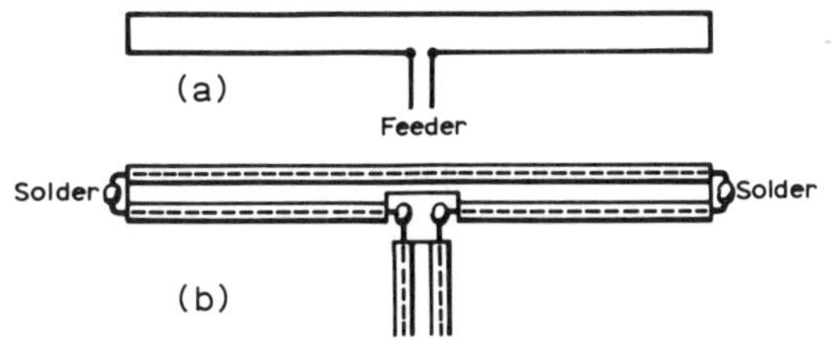

Fig 7.14. (a) Folded dipole. (b) Construction of a folded dipole from 300Ω ribbon feeder

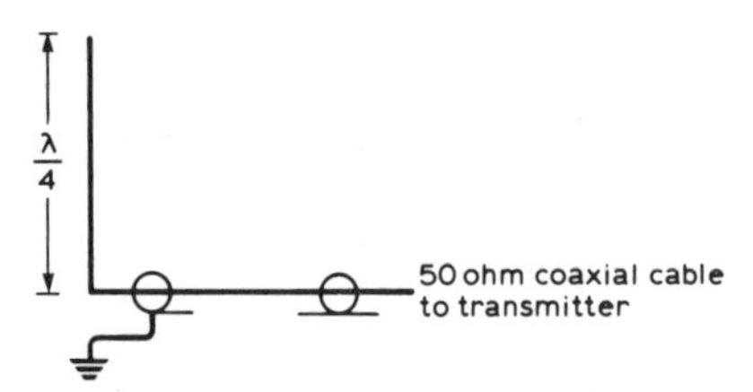

Fig 7.15. λ/4 vertical antenna

The achievement of a satisfactory earth presents the major difficulty. A single earth rod, say 2m long, is unlikely to be satisfactory unless the soil has exceptional conductivity. Two or three such rods bonded together close to the bottom of the radiator may reduce the earth resistance.

Earthing problems with a vertical antenna may be virtually eliminated by erecting it over a perfectly-conducting surface, eg a large sheet of copper. This is known as a 'ground-plane antenna', but is only realisable at VHF (eg a λ/4 radiator mounted on a car roof).

In practice a satisfactory ground plane may be made by laying four to six radial wires about λ/4 long on the surface of the ground (they can be buried a few centimetres below the surface if more convenient). Alternatively, the radiator may be mounted at the top of a mast which uses the ground-plane radials as guy wires, insulators being introduced at the appropriate points as shown in Fig 7.16. A ground plane erected in this manner does in fact present a better match to 50Ω coaxial cable than does the conventional ground plane.

Traps may be inserted in a vertical antenna to enable it to be used on more than one band.

The end-fed antenna

This is probably the simplest antenna of all as it consists of a length of wire brought from the highest point available direct to the transmitter output. The wire can be straight but good results are often obtained with quite sharp bends in the run of the wire.

Optimum results are obtained with resonant lengths, a 40m long end-fed antenna operates on bands from 3.5MHz to 28MHz, while an 80m length enables 1.8MHz to be used. This arrangement, although often used, is liable to create breakthrough problems (Chapter 9) as the end of the antenna

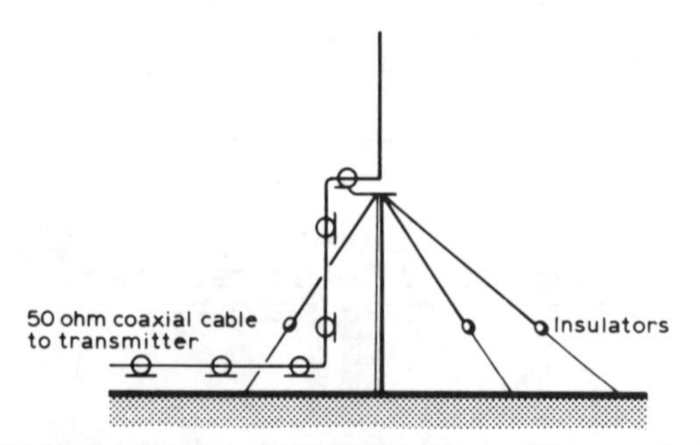

Fig 7.16. Ground-plane antenna mounted on a mast

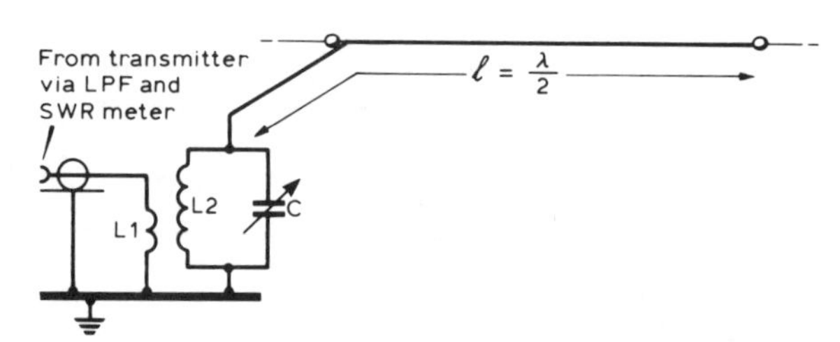

Fig 7.17. End-fed antenna

(ie a high-voltage point) is brought into the house; if the transmitter is not at ground level there may be significant unwanted radiation from the long earth lead. The use of an antenna tuning unit as shown in Fig 7.17 is advisable. The ATU should be preceded by a low-pass filter and SWR meter as shown in Fig 7.10.

The three-element beam

The Yagi antenna consisting of a radiator and two parasitic elements (a reflector and one director) mounted on a suitable tower with provision for rotating it is known as a 'three-element beam'. The radiator is a dipole which must be about 10m in overall length for operation at 14MHz; thus physical size normally dictates that 14MHz is the lowest frequency for which the rotating Yagi is used.

The addition of a director and a reflector to the normal dipole has the effect of reducing the centre impedance to the order of 20Ω. Therefore in order to feed it with 70Ω coaxial cable the impedance must be increased. This can be achieved by folding the radiator which increases the impedance to about 80Ω or using some form of impedance-matching transformer between the feeder and the antenna.

The actual spacings of the reflector and director relative to the radiator have a significant effect on the characteristics of the system, such as the gain compared with a simple dipole, the front-to-back ratio and the amount by which the impedance is reduced.

Commercial three-element beams are widely used. These use traps for operation on 28MHz, 21MHz and 14MHz. As a result of the different spacing in terms of operating wavelength at these three frequencies, the change in impedance is not excessive, and feeding with 50Ω coaxial cable is an acceptable compromise.

The quad antenna

The quad antenna consists of a square loop of wire as shown in Fig 7.18. The side of the loop is approximately λ/4 in length and it is normally fed with 75Ω coaxial cable at the point shown. The loop can be mounted with a diagonal vertical and fed at the bottom corner; in either case the feed point is a current maximum and the performance is identical. In the configuration shown, the polarisation is horizontal.

Parasitic elements may be added to form a beam antenna. The most popular is a radiator plus reflector (the two-element quad), but one or more directors may be added, depending on

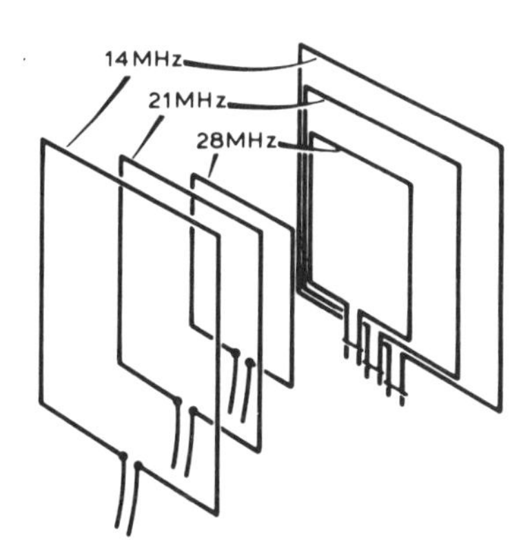

Fig 7.18. A three-band nest of two-element quads (radiator and reflector) maintaining optimum spacing for each band

the frequency. Spacings between the radiator and parasitic elements are similar to the Yagi. Commonly, quads for 28MHz, 21MHz and 14MHz are assembled on the same mounting and rotating system and fed by 50Ω cable. In order to maintain the optimum spacing between the elements, the radiators, reflectors and directors for each band cannot be in the same vertical plane (see Fig 7.18).

The quad is made up of wire supported on light-weight spreaders of bamboo or glass fibre.

Front-to-back ratio and SWR are optimised on each band by adjustment of the tuning stub on the reflectors, but alternatively a reasonable compromise is often obtained by eliminating the stub and making the reflector about 3% greater than the radiator in length.

Commercial data suggest that the quad may give a slightly higher gain than the Yagi and a noticeably better front-to-back ratio. Its smaller turning radius is also often advantageous.

Low-frequency antennas

An effective resonant antenna for the LF bands, particularly 1.8MHz, requires a large space. Shorter lengths are often used tuned against earth in what is known as the 'Marconi-type antenna', as shown in Fig 7.19; the slightly different tuning arrangement should be noted. A common arrangement is shown in Fig 7.19. The earth should be a short connection to an earth spike and the use of the mains earth or the water-pipe system should be avoided.

VHF and UHF antennas

Resonant antennas, as discussed in this chapter, are in general applicable to all amateur bands up to 1300MHz.

The Yagi, and to a lesser extent, the quad, are widely used at VHF and UHF because the much shorter physical length of a half-wavelength means that more elements occupy less space and so antennas of appreciably higher gain than at HF are possible. Several such antennas may be stacked, provided they are correctly matched and phased, to provide even higher gain.

At even higher frequencies, parasitic elements are replaced

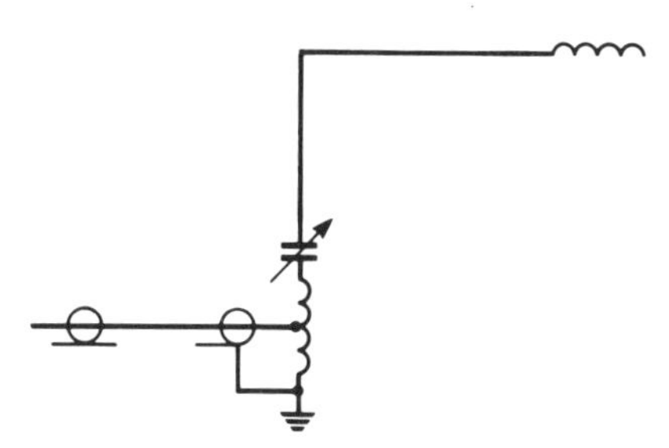

Fig 7.19. Marconi antenna for use at 1.8MHz; additional loading at outer end is useful if length is less than $\lambda/4$

by parabolic-shaped reflectors or 'dishes'; however, these are beyond the scope of the RAE.

Loading of an antenna

Shortness in length of a resonant antenna may be compensated to a certain extent by the addition of a small amount of inductance, for instance as shown in Fig 7.19. Another example is the effect of the trap inductance when the trap dipole (Fig 7.13) is used on the 3.5MHz band. This artifice is commonly used on whip antennas for mobile use and may allow a beam antenna to be made to fit into a smaller than normal space. The process of loading an antenna may degrade its properties.

Power radiated from an antenna

Effective radiated power (ERP)

The ERP in a given direction is the product of the power supplied to the antenna and the gain of the antenna relative to a half-wave dipole in that direction.

Effective isotropically radiated power (EIRP)

The product of the power supplied to the antenna and its gain, in a given direction, relative to an isotropic (ie omnidirectional) antenna. By convention, ERP is used below 1GHz and EIRP above 1GHz.

In assessing the power supplied to the antenna, the loss in the feeder system must be taken into account. This is negligible in the lower-frequency bands but significant at UHF and above.

The calculation of the field strength produced at a point distant from the transmitting antenna under practical conditions is difficult. This is because the signal is reflected and absorbed by surrounding objects, particularly those close to the transmitting antenna and especially the ground, and, of course, the ionosphere. This is not generally important in amateur radio, but from the aspect of EMC (see Chapter 9) it is often useful to obtain some indication of the likely field strength at quite short distances in order to assess the likelihood of breakthrough of your signal to other equipment in the immediate vicinity.

A formula derived from free-space propagation theory can be used to calculate field strength under these conditions, ie

$$e = \frac{7.02\sqrt{\text{ERP}}}{d}$$

where *e* is the field strength (peak) in volts per metre and *d* is distance from transmitter in metres.

For example, if your transmitter has an ERP of 400W, what is the field strength at a distance of 100m? From the above expression

$$e = \frac{7.02\sqrt{\text{ERP}}}{d}$$

$$= \frac{7.02\sqrt{400}}{100} = \frac{7.02}{5}$$

$$= 1.4 \text{ volts/metre}$$

Receiving antennas

It is normal practice for the transmitting antenna to be used for reception. Should a separate receiving antenna be required, 12–18m of wire erected in the clear normally gives good results on all bands. An antenna tuning unit is often advantageous.

CHAPTER 8

Transmitter interference

This chapter is primarily concerned with how interference is caused to fellow radio amateurs and to other services using radio frequencies. It should be read in conjunction with Chapter 9 which deals with electromagnetic compatibility, good station design and methods of minimising interference in the home and to immediate neighbours, and with Chapter 4 which contains useful information about transmitter design.

Interference caused by amateur transmissions may be classified broadly into two groups. In one, interference is to users of immediately adjacent frequencies. In the other, frequencies much further away are affected. In either case, as interference may be caused at great distances from the transmitter, the amateur has a heavy responsibility to ensure that his transmissions do not attract criticism or, worse still, break the conditions of the amateur licence. It is important to realise that the fact that a transmitter is of commercial manufacture does not necessarily mean that is free of interference problems. The following examples of transmitter defects and shortcomings apply both to home-constructed and commercial equipment.

Frequency instability

The frequency of a VFO will vary if the mechanical stability of the circuit and its components is poor, as explained in Chapter 4. Apart from this, two other common forms of frequency instability are 'drift' and 'chirp'.

'Drift' is a gradual change of frequency which may occur on any mode of transmission. It is most objectionable on SSB where even a small change of the transmitting frequency requires the receiver to be retuned. After having established contact on one frequency, two stations may slowly drift away from each other, ending up several kilohertz apart. This occupies an unnecessary space in the RF spectrum. Severe drift may result in transmissions ending up outside the edges of the amateur bands. It usually occurs because certain components in an oscillator circuit change value as the transmitter warms up. A variable frequency oscillator (VFO) is worst affected. Although a VFO DC supply voltage is invariably stabilised, this too may vary slightly with temperature, causing further unwanted frequency changes.

'Chirp' is a characteristic of A1A mode morse (CW) signals. It is a rapid change of frequency which occurs for a fraction of a second after the morse key is depressed. One cause is a poorly regulated VFO DC voltage supply which falls slightly because the transmitter draws more current from the power unit during key-down conditions. Another cause of chirp is unwanted feedback of the transmitted signal into the VFO circuit. This is minimised by good screening and filtering of the power leads supplying the VFO. The effects of feedback are most noticeable if the VFO and transmit frequencies are the same, so it is common practice to mix or to multiply the VFO signal to obtain the required transmitter output frequency. The VFO stage should be followed by a buffer amplifier. This stops RF signals entering the output of the VFO. It also prevents chirp which may occur if the loading on the VFO changes between key-up and key-down conditions.

Transmissions which occupy excessive bandwidth

On the crowded amateur bands, the amateur must ensure that he does not make excessive use of the space available by limiting the bandwidth of his transmissions to the minimum necessary for effective communication.

A signal which consists only of a single carrier wave occupies no bandwidth. It also conveys no information. The information is conveyed by modulating the carrier, frequency modulation (FM) and amplitude modulation (AM) being the most popular modes used by amateurs. Morse and single-sideband transmissions are both forms of AM. Modulating the carrier produces sidebands (see Chapter 4) and it is really these which contain the information. The complete transmission therefore occupies a band of frequencies.

Speech

Most amateur speech transmissions use single sideband (SSB, mode J3E) or FM (mode F3E). Ordinary AM (A3E mode) is rarely used these days. All may occupy a considerable bandwidth unless certain precautions are taken. Any attempt to transmit high-quality audio (which contains frequencies up to 15kHz) will result in the use of excessive bandwidth, and result in 'splatter' across adjacent channels. A frequency range from 300Hz to a top limit of 2.5 or 3kHz provides adequate intelligibility for communication purposes.

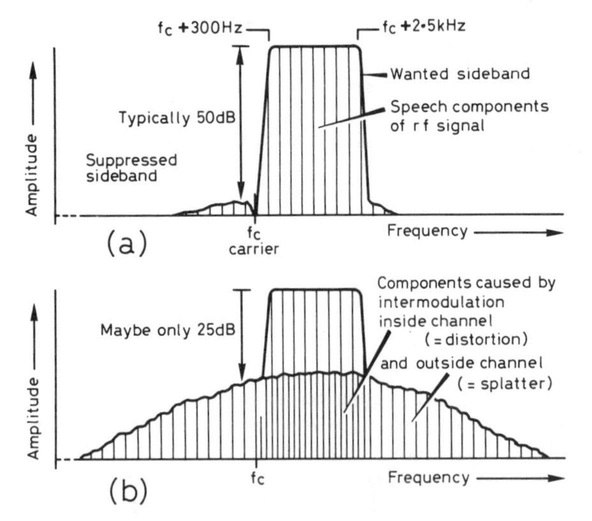

Fig 8.1. Spectrum of an SSB signal (a) when first generated; (b) after passing through an over-driven 'linear' amplifier

In AM transmissions, the bandwidth which should be occupied is twice the highest audio frequency, eg 2 × 2.5kHz = 5kHz. The transmitter audio stages should be designed to have an overall low-pass characteristic with a sharp cut-off above 2.5kHz. However, severe splatter will result if the carrier is 'over-modulated'. This occurs when too much audio signal is applied to the carrier, resulting in breaks during the modulation (see Fig 4.12). Too little audio makes the signal appear weak, so to ensure maximum intelligibility the modulation should be kept close to 100%. It is advantageous to use some form of automatic audio compression or limiting circuit which, when adjusted correctly, ensures a consistently high level of modulation but prevents over-modulation.

In SSB transmissions all the power of the signal is concentrated into one sideband only. Ideally, the unwanted sideband and the carrier would be totally suppressed and transmitted bandwidth that of the original audio modulating signal (ie somewhat less than half that required for AM). In practice the suppression is typically 50dB. The spectrum of an SSB signal is shown in Fig 8.1(a). It can be seen that the signal will occupy excessive space if the sideband suppression is insufficient. Poor carrier suppression may cause annoying heterodyne whistles to those trying to operate on adjacent frequencies.

Unlike AM (where it is normal practice to modulate the transmitter output stage), SSB is usually generated in an early stage and the required transmitter power obtained by using a 'linear' amplifier which should cause negligible distortion. However, if the amplifier is over-driven in order to obtain a power output of which it is not really capable, serious distortion will result. This causes intermodulation to occur. New, unwanted signals are generated, both inside the transmitted channel (resulting in audible distortion) and outside (resulting in splatter across adjacent frequencies). The signal therefore occupies a greater bandwidth than before amplification, as shown in Fig 8.1(b). Over-driving can be prevented by the use of compression or limiting circuits, and these may operate either at audio or at radio frequencies.

When using FM, the audio signal deviates the frequency of the transmission from the nominal carrier frequency. This generates sidebands in a manner somewhat similar to AM. However, on FM an audio tone of, say, 2.5kHz creates sidebands not only at 2.5kHz, but also at harmonics of 2.5kHz (ie 5, 7.5, 10kHz etc). In theory these stretch to infinity each side of the carrier. In practice FM is not quite as antisocial as it appears, because the level of the additional sidebands depends greatly on the level of the modulating signal. They are negligibly small if the modulation index does not exceed about 0.6. The bandwidth is then comparable with an AM transmission. To achieve this, it is essential that the audio bandwidth and amplitude be kept within well-defined limits (probably even more so than with AM and SSB transmissions) and audio compression or limiting is used in almost every FM transmitter.

For amateur narrow-band FM (NBFM) the maximum deviation should not exceed ±2.5kHz. As well as causing severe splatter over adjacent frequencies, a very noticeable characteristic of over-deviated FM signals is that, even when correctly tuned in, the received audio sounds distorted on modulation peaks, and the receiver S-meter kicks noticeably downward. Even when the deviation is correct, the higher sidebands are not entirely negligible. To prevent out-of-band radiation, FM transmissions must not be less than 10kHz from the edge of the amateur band.

Morse (CW, A1A mode)

Morse is usually transmitted by on-off keying of an otherwise continuous wave, and is usually referred to by amateurs as 'CW'. Methods of keying vary. In high-power transmitters, the supply current or a bias voltage in a low-power stage is keyed. In low-power transmitters the PA stage itself may be keyed. Whichever method is used, care must be taken to prevent key-clicks being produced.

Only a relatively narrow bandwidth is needed for morse – a few hundred hertz at the most at normal keying speeds. However, if the keying turns the transmitter on and off instantaneously, a signal with a very wide bandwidth is produced.

Fig 8.2 shows the transmission of the morse character 'A'. The instantaneous switching of the signal from zero to full power and back to zero is equivalent to 100% amplitude modulation by a rectangular-waveform audio signal. The sharp edges of the waveform produce sidebands which, on adjacent frequencies, are audible as clicks each time the key is pressed or released. The sharper the edge, the greater is the bandwidth occupied by the sidebands. Fig 8.2(c) shows the waveform without filtering, where the clicks may be audible over 100kHz away from the signal carrier frequency. However, although the keying may sound 'hard' (and probably rather tiring after some time), the disruption caused to users of nearby frequencies may not be realised by listening to the signal itself.

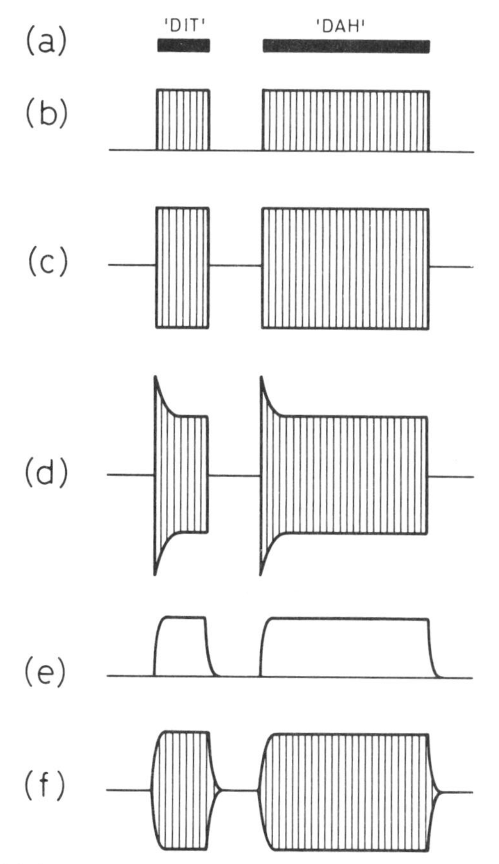

Fig 8.2. Comparison of keying waveforms with and without filtering. (a) The morse letter 'A'. (b) The supply current to the keyed stage without filtering. (c) The resulting RF signal envelope. Note the sharp edges. (d) The RF envelope when power supply regulation is poor. (e) The slowing of the rise and fall of the keying current using a key-click filter (see Fig 8.3). (f) The resulting RF signal envelope with soft edges and hence minimal key clicks

The situation is made worse if the transmitter PA power supply suffers from poor regulation and the DC voltage rises considerably during the key-up period. There is a surge of power at the instant the key is pressed, as shown in Fig 8.2(d). This accentuates the key-clicks and gives the received signal an unpleasant, thumping characteristic.

Key-clicks are suppressed by slowing down the rise and fall of the keying waveform, as shown in Fig 8.2(e). This in turn softens the transitions of the power output waveform, reducing the sideband spread and making the signal more pleasant to listen to. Fig 8.3 shows a simple filter circuit suitable for use where the key interrupts the DC current flow in a transmitter stage. When the key closes, the rise of current is slowed down by the inductor in series with the key. When the key opens, the fall of current is slowed down by the charging of the capacitor across the key. The resistor is necessary to prevent the rapid discharge of the capacitor at the instant when the key closes. Without it, sparking would occur at the key contacts.

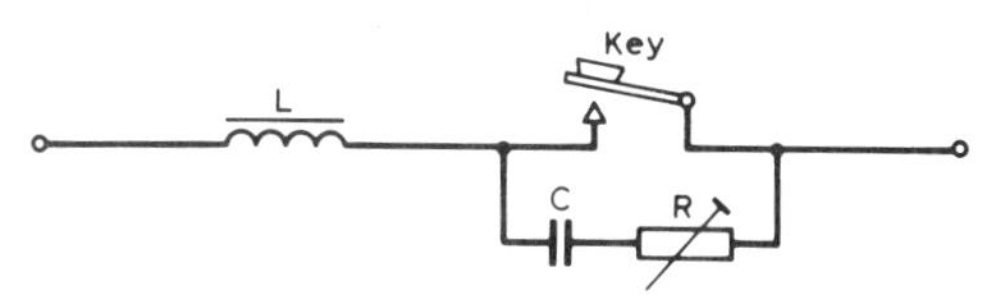

Fig 8.3. Typical key-click filter. L serves to prevent too rapid a rise of current. C, charging through R, serves to continue flow of current momentarily when key contacts open. Typical values: L = 0.01 to 0.1H, C = 0.01 to 0.1μF, R = 10 to 100Ω

Unwanted transmitter outputs

Harmonics

The closing down of the UK VHF 405-line TV service has removed what was probably the most common reason for complaints of amateur interference, ie harmonic problems with HF band transmissions (both radiated by the transmitter and self-generated in the TV set itself). However, the growth of mobile and other radio services on the vacated TV (and other) frequencies means that high standards for unwanted transmitter outputs must be maintained. There is a good chance that an amateur will live close to users of VHF and UHF services (even next door), and very low levels of unwanted outputs could cause severe interference. Harmonics which fall below about 30MHz may not be as troublesome because users of HF are relatively few and hence likely to be more remote. With the relatively high levels of HF background noise and interference, low levels of unwanted signals will probably go unnoticed.

Without attention to filtering there is a progressive (if somewhat erratic) drop in the level of the higher harmonics. Fig 8.4 shows the comparison of harmonic levels of a 10MHz transmitting station, illustrating the improvement obtained by good station design. Harmonics from the lower frequency amateur bands may fall within a higher band (eg 2×3.53MHz = 7.06MHz, thus interfering with other amateurs) or outside the HF end (eg 2 × 3.7MHz = 7.4MHz, thus interfering with other services). Interference may involve TV or FM radio, leading to complaints from neighbours, eg 5 × 144MHz = 720MHz. This is within TV Channel 52. Further examples are given in Chapter 9.

Harmonics occur because amplifiers are to some extent non-linear, ie they distort the signal waveform. They are produced mainly in the high-power stages in a transmitter, especially those which use Class C amplification. This may be the most efficient, but it is highly non-linear. Even the relatively linear Class A and Class B amplifiers used in SSB transmitters are by no means harmonic-free.

Most transmitters with a tuned PA stage use 'pi-network' output tuning (or some form thereof, such as 'L-pi'). Fig 4.9 is a typical example. This acts as a built-in low-pass filter and greatly reduces harmonic levels. In those transmitters where wideband (untuned) power amplification is used, different bandpass or low-pass filters are often used (being selected as required by the transmitter bandswitch).

Provided that the transmitter is fully screened and all the leads entering or leaving the case are fully filtered and

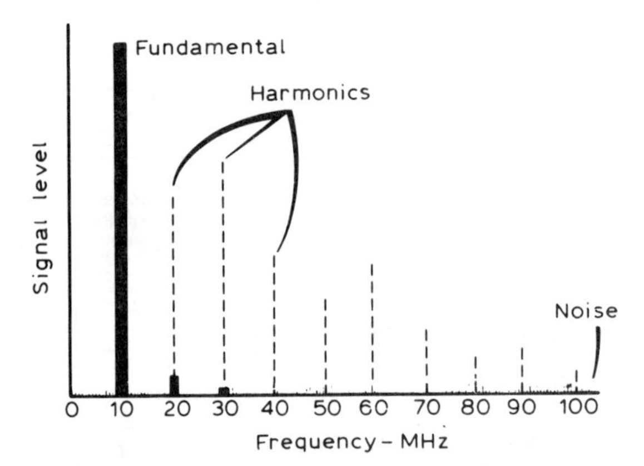

Fig 8.4. Example of harmonics received from a 10MHz station. With poor station design, harmonic levels may be excessive (shown dotted). With good design it will be difficult to detect even the 2nd or 3rd harmonics above the received general noise level

decoupled, then harmonics should be no problem. As the only way out for RF signals is via the antenna socket, then any harmonics can be removed by a suitable filter fitted in the feed to the antenna system (see Fig 9.6). However, filtering is made easier if the production of harmonics is minimised in the first place by (for example) avoiding unnecessarily high drive levels in power amplifiers. Stages carrying RF signals should be individually screened and their power feeds well decoupled in order to prevent RF signals and harmonics from being carried around inside the transmitter via the internal wiring. A good-quality metal case which has no large holes, gaps or slots (which will allow RF leakage) will then complete the task of screening the transmitter.

Spurious oscillations

Amplifier stages have an unfortunate tendency to oscillate if there is sufficient (unintentional) feedback from output to input. The presence of such spurious oscillations in some part of a transmitter circuit may result in radiation on essentially unpredictable frequencies in addition to – and possibly as powerful as – the wanted frequency. This may result in severe interference to other radio users over a wide area.

In many respects the techniques of filtering, decoupling, screening etc, intended to minimise harmonics, also help to prevent spurious oscillations. However, some additional precautions may be necessary. Note that if particularly violent oscillations occur, some transistors (especially the more expensive devices used at VHF and UHF) may be instantly destroyed. Take great care when investigating oscillations. It is wise to operate the suspected stage at a reduced voltage and with some current-limiting resistance in series with the DC supply.

'Self-oscillation' may occur at or near the working frequency of a stage, especially if both input and output are tuned to the same frequency. It may occur only when no RF input drive is present, eg between words on SSB or with key-up on CW. Its frequency will be fairly unstable and the transmitter may continue to give a noticeable RF output when the power level should read zero. Conversely, the oscillation may occur only when the stage is being driven. The result may be burbly, splashy speech and a rough CW note, with chirps and key-clicks. The cure for this problem is to minimise the feedback by ensuring that input and output circuits are screened from each other. DC feeds may also provide another feedback path and should be adequately decoupled. There may be enough feedback through the amplifying device itself to cause oscillation, in which case the stage will have to be 'neutralised'.

Oscillations may also occur at relatively low frequencies and sometimes modulate the RF signal, appearing as sidebands (often multiple) either side of the wanted output frequency. They may be the result of instability in audio stages or in regulated DC power supplies (often at supersonic frequencies) or by the chance resonances of RF chokes and decoupling capacitors, often in the DC feeds to certain stages. For example, VHF transistors often have a very high gain at much lower frequencies, and care must be taken to ensure that the decoupling of the DC supplies is effective at these low frequencies. Fig 8.5 shows how a 'hidden' parallel resonant circuit exists. To prevent the danger of oscillations the resonance must be damped, either by using a very-low-Q (ie 'lossy') choke for L1 or by adding R1 and C4 to the existing circuit.

VHF and UHF oscillations may occur in HF transmitters because of a poor choice of circuit layout, and the type and style of components used. This often results in capacitor leads, earth returns and other interconnections which are rather long, and are therefore somewhat inductive. Coils also may have long leads which are not part of the winding. These small, unintentional inductors have no adverse effect at HF, but form resonances at much higher frequencies with the stray capacitances in the circuit. Indeed, these 'parasitic' resonances may affect even well-laid-out circuits where lead lengths cannot be shortened and the correct style of components are used. As with low-frequency resonances, the cure is to make

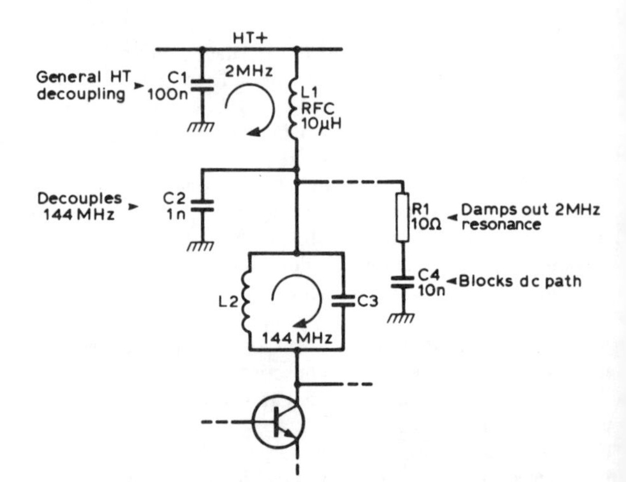

Fig 8.5. Suppression of low-frequency spurious resonance

these resonances well damped. Low-Q VHF chokes (eg coils wound on resistors or on lossy ferrite beads) may be added at suitable points in the circuit to break up the parasitic resonances.

Mixer products

Many transmitters use mixing processes to obtain the required output frequency – especially for SSB and VFO-controlled VHF transmissions. Mixers (frequency changers or converters) produce not only the wanted sum or difference frequencies (see Chapter 2) but also a whole range of undesirable signals. Most significant are harmonics of the local oscillator injection and harmonic mixing products, ie outputs created when the self-generated oscillator harmonics mix with the input signal to produce a large number of sum and difference signals.

For example, an SSB signal generated on 9MHz may be converted for transmission on 14MHz by using a VFO on 5MHz (9MHz + 5MHz = 14MHz). Unfortunately the mixer is likely to generate the third harmonic of the VFO on 3 × 5MHz, ie 15MHz. It may also generate the fifth harmonic, which will mix with the 9MHz and appear at 16MHz (5 × 5MHz – 9MHz = 16MHz). Considerable filtering will be needed to ensure that these unwanted products are sufficiently reduced in level by the time that they reach the transmitter output. Balanced and double-balanced mixers help to reduce such problems because if the balance is perfect, the oscillator signal and its harmonics do not appear at the mixer output.

Unwanted mixer products may be minimised as follows.

1. Choose signal generation and oscillator frequencies carefully so that unwanted outputs are remote from the wanted output. This simplifies subsequent filtering.
2. Do not over-drive mixers either with input signal or oscillator injection. This reduces harmonic generation and harmonic mixing.
3. Use balanced and double-balanced mixers to reduce the number of unwanted outputs.

Synthesiser problems

Modern commercial transmitters rarely use the low-frequency VFO and frequency multiplier system. Instead, a phase-locked loop (PLL) frequency synthesiser replaces the VFO and may operate at very much higher frequencies where a VFO would be far too unstable.

Synthesisers contain a crystal oscillator (typically 1 to 15MHz), voltage-controlled VFO (VCO) and digital divider circuits. There are often additional circuits such as frequency multipliers and mixers. The synthesiser circuit must be adequately screened and filtered to ensure that it does not radiate into or pick up from other parts of the transmitter. This could result in the transmission of signals on some very unexpected frequencies. An important failsafe design feature of a synthesiser is that the transmission must be inhibited if the VCO has not locked to the reference frequency. This prevents the radiation of a very unstable signal (albeit only momentarily, under normal working conditions).

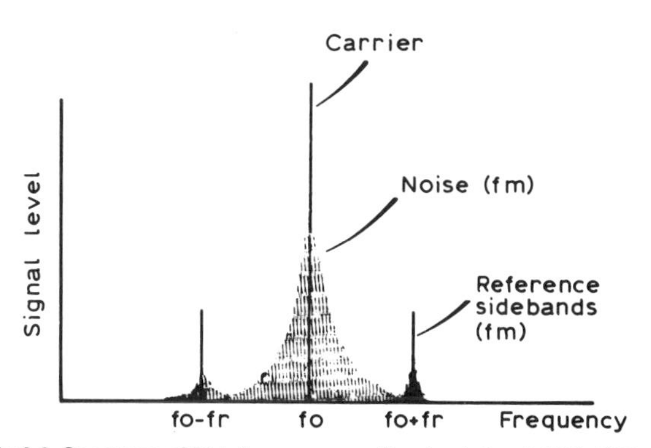

Fig 8.6. Spectrum of PLL frequency synthesised signal. With VHF and UHF transmitters f_r is often equal to the channel spacing, and sidebands $f_0 - f_r$ and $f_0 + f_r$ fall on adjacent channels. Good design minimises both noise and reference sidebands

Even the wanted output signal can be troublesome if the design does not minimise unwanted frequency modulation by noise (inherently greater than with a VFO or crystal oscillator) or by the reference frequency. This is often the smallest frequency step available and in channelised equipment (as used at VHF and UHF) is typically 12.5kHz or 5kHz. Fig 8.6 shows the spectrum of an otherwise unmodulated signal which could give rise to complaints of 'hash' or 'whine' on adjacent frequencies. Note that when modulated, the whole signal (and not just the carrier) will carry modulation sidebands, giving the distinct impression that the signal is being over-modulated.

Frequency measurement

It is essential that no transmissions be made outside the amateur bands, and signals should be located sufficiently inside band edges to allow for the spread of sidebands. The requirements of the DTI for frequency checking and measuring equipment are dealt with in Chapter 10.

An inductively coupled absorption wavemeter is invaluable as a test instrument during the construction of transmitters, but for day-to-day operation of the station it is more convenient to use a tuneable 'field strength' meter (misnamed). This may consist of an absorption wavemeter equipped with a short whip antenna, a detector circuit and a microammeter, and is used to check that the transmission is on the intended band. This avoids the possibility of tuning a transmitter to the wrong band which can happen, for example, if the bandswitch is set for 7MHz but the PA is inadvertently resonated on 14MHz. This error may not be immediately apparent on the station receiver.

If the amateur uses a separate transmitter and receiver it is useful to have available a crystal calibrator, preferably built into the receiver. This will enable the receiver tuning accuracy to be checked at frequent intervals. A good-quality crystal and a few low-cost digital integrated circuits can provide frequency calibration markers typically at intervals of every 10MHz, 1MHz, 100kHz, and 10kHz. The accuracy of the frequency of the amateur transmission (and indeed any other signal) can then be checked against the markers.

If the amateur uses a transceiver it is not possible to transmit and receive at the same time. In these circumstances only the receiver can be calibrated, and further checks should be made to verify that the transmitting and receiving frequencies really are the same. Where the equipment incorporates a digital frequency display it is unwise to assume that this is absolutely accurate. The transmission frequency in particular should be checked, and due allowance made for possible inaccuracies when operating near band edges. Any equipment used to measure frequency (crystal calibrator, heterodyne wavemeter, digital frequency counter etc) should be checked regularly and, if necessary, adjusted. The Standard Frequency Transmissions radiated on 2.5, 5 and 10MHz by various transmitters (in the UK, by MSF at Rugby) and the 198kHz BBC Radio 4 transmissions are readily available standards for calibration purposes.

CHAPTER 9

Electromagnetic compatibility

With the increasingly intensive use of the RF spectrum and growth in consumer electronics, it is becoming more important that equipment be designed with electromagnetic compatibility (EMC) in mind. In the past EMC used to be a specialist subject but it has now become a topic that affects everyone who uses electronic equipment. It is therefore important that everyone has an understanding of their responsibilities, the problems that can occur and ways of curing them.

Electromagnetic compatibility is defined as 'the ability of a device, equipment or system to function satisfactorily in its electromagnetic environment without introducing intolerable electromagnetic disturbances to anything in that environment'. It has two main aims: to prevent pollution of the RF spectrum so that it can be used for radio communications, and to ensure that equipment which does not use the RF spectrum is able to operate correctly in the presence of the RF fields that exist in the environment. It can cover both radiated signals and those conducted down cables, eg transients on the mains. Examples of the pollution include interference from arcing thermostats, noise from computers, or spurious emissions from transmitters. An example of inadequate immunity to RF fields is when hi-fi equipment detects radio transmissions which can be heard on the system's loudspeakers.

The radio amateur and EMC

In the UK, the cessation of the VHF (405 line) TV transmissions has resulted in a great reduction of the interference which used to occur because of the harmonic relationship between most of the HF amateur bands and the TV channels (especially those in Band 1 – 40 to 70MHz). These days, interference is rarely due to a serious technical shortcoming in the transmitter. It is more likely to be due to the inability of the equipment suffering the interference to withstand the presence of the strong amateur signal. However, it must be remembered that the old TV frequencies have been allocated to other radio services, and standards for minimising harmonics and other unwanted transmitter outputs must not be relaxed.

It is useful to be able to picture where the amateur bands lie in relation to the frequencies used by other services. Fig 9.1 shows some important allocations. If either the transmitted or received signals require additional filtering, it gives an indication of what type of filter might be suitable.

It is the purpose of a transmitting station to generate and radiate RF fields and it is inevitable that these fields will be fairly strong in the immediate vicinity of the station. Much of the modern consumer electronic equipment which is not concerned in any way with receiving radio signals has not been designed to operate in this environment and responds to these RF fields, resulting in breakthrough problems. Problems can also occur in the reverse direction; the consumer equipment can radiate noise and other unwanted emissions which can interfere with the reception of signals. These two types of problem often go hand-in-hand, and the design techniques which solve one problem will often also solve the other.

The use of transmitters and receivers in all areas of life is growing steadily and so these problems have to be solved. The best solution is for immunity to be designed into the equipment at manufacture, as it is much harder to correct retrospectively. Improved standards are under discussion. However, dramatic improvements can be made to existing equipment by the simple filtering measures which are described later in this chapter.

The amateur can help to reduce the impact of the problem by careful design of his station. In particular, the antenna should be sited to minimise the level of RF fields in neighbouring property, while maximising the useful radiation which clears the immediate vicinity and goes on to reach the distant receiving stations.

It must be emphasised that EMC problems *can* be solved if all parties co-operate.

Curing EMC problems

Filters

Fig 2.17 shows the basic frequency responses of two commonly used filters: the low-pass and the high-pass. Fig 9.2 shows the characteristics of two others: the bandpass and the bandstop. The purpose of a filter is to pass the wanted frequencies and reject the unwanted. The type most suitable for a particular application may be determined from the spectrum chart shown in Fig 9.1. For example, to prevent the harmonics of a multi-band HF transmitter (1.81 to 29.7MHz)

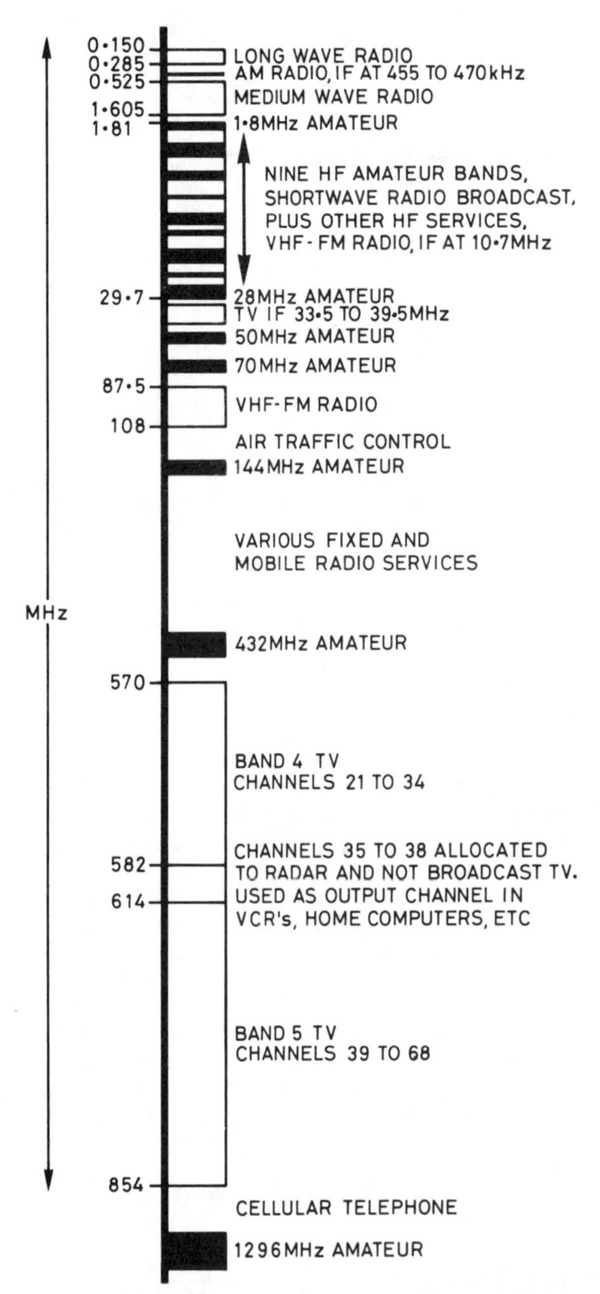

Fig 9.1. Spectrum chart indicating the frequencies used by amateurs in relation to those used by TV and radio broadcast receivers

from causing interference to broadcast VHF-FM radio, UHF TV and other VHF/UHF services, a low-pass filter (LPF) with a cut-off frequency of 30MHz is often inserted in the feed to the transmitting antenna.

Where a UHF TV set is experiencing breakthrough from, say, strong local 144MHz transmissions, a high-pass filter could be fitted in the receiving antenna coaxial downlead (usually at the antenna 'input' socket). However, for problems with HF transmissions, a 'braid-breaker' is more appropriate. This is discussed later.

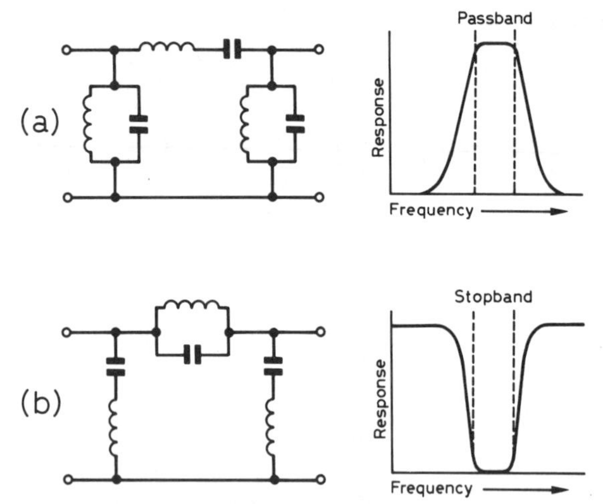

Fig 9.2. Examples of circuits and frequency response curves of (a) a bandpass filter, (b) a bandstop filter

Bandpass and bandstop filters also have their applications. Where a single-band transmitter is used, a bandpass filter in the antenna feed will pass the wanted frequency, but remove unwanted signals both above and below it. A bandstop filter does the opposite. It could be used in the radio or TV downlead to protect the set from the strong signal radiated by the transmitter because it operates over a relatively narrow frequency range.

All filters introduce some insertion loss at the frequency of the wanted signal, and it is an important design parameter that this loss should be acceptably low (ie less than 1dB). The sharpness of the transition between the passband and the stopband (and how well it rejects an unwanted signal which is close to the wanted) depends upon the complexity of the circuit. For example, if the simple HPF circuit of Fig 2.17(b) had a cut off frequency (f_c) of 470MHz (for passing UHF TV signals), it would give little protection against transmissions on 432MHz.

Another important filter parameter is its characteristic impedance. A filter inserted in a 50Ω feeder should, inside its passband, appear simply as another piece of 50Ω feeder, ie it should match the feeder characteristic impedance. The match may not be perfect, but a good filter will exhibit a low SWR, ie 1.2:1 or less. A mismatched filter will have a high SWR, a high insertion loss and often a degraded stopband.

Ferrite ring (toroidal) chokes

The ferrite ring choke is one electrical component which finds frequent application in the fight against EMC problems. In the transmitting station, it may be used to block the flow of RF currents (for example, into the mains wiring or on the braid of the antenna feeder) which, if unsuppressed, would increase

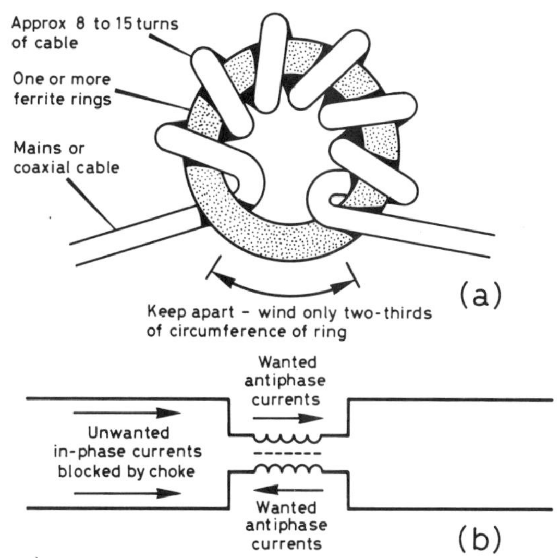

Fig 9.3. (a) Winding details of a ferrite ring choke for use as a mains filter or as a braid-breaker in a TV or FM radio coaxial downlead. (b) How it behaves electrically

the local field strength. It may also be similarly used with the affected equipment in order to block RF currents which would otherwise flow along leads and into the equipment.

Fig 9.3 shows how a choke should be constructed. The main requirement is that it have sufficient inductance at the interfering frequency. However, to maximise the self-resonant frequency, an excessive number of turns should be avoided, and the start and finish of the windings kept apart. A correctly made choke is usually effective over a wide frequency range.

Unwanted signals which are picked up on leads usually flow together (ie 'in phase' or 'longitudinally') along all the wires comprising the lead. The wanted signal currents flow in antiphase, and are not affected by the choke. It can therefore be used as a 'braid-breaker' which blocks interference picked up on the braid of a TV or FM radio coaxial downlead, as a mains filter on a twin or three-wire lead, on audio and video screened or balanced twin leads etc.

There are alternative forms of the choke, such as a two-part core and clamp, a long ferrite rod (with many turns) or even a non-inductive former (eg a plastic tube), provided that sufficient inductance can be obtained.

Mains filters

The simplest, but by no means least effective, mains filter is the ferrite ring choke of Fig 9.3. If it is sufficiently thin and flexible, the mains lead itself can be wound on the ring. With more rigid cable, fewer turns may be used, but with several cores stacked together to maintain the inductance.

Certain commercial units (electric motor interference filters, 'mains conditioners' for computers etc) are designed primarily to suppress transients between the live and neutral conductors. Many have a common earth connection between the input and output terminals through which interfering currents can pass without being filtered. These can be used only where the earth continuation is not required, or where it can be choked separately, using a wire gauge which is sufficient to carry full fault-condition currents.

Braid-breakers

The purpose of a braid-breaker is to block unwanted currents which result from RF pick-up on the braid (outer screen) of a coaxial cable (eg a TV antenna downlead). Again, the ferrite ring choke of Fig 9.3 is ideal for this purpose, wound using the cable itself, or, if this is too rigid, with miniature coaxial cable fitted with suitable mating connectors. The wanted signals picked up by the antenna propagate in the downlead in antiphase, and are not affected by the inductance of the choke. The insertion loss is negligible, even at UHF TV frequencies.

Fig 9.4. Braid-breaker using a 1:1 ratio RF transformer which can only pass the wanted antiphase currents. R is a static discharge path – 10kΩ or greater

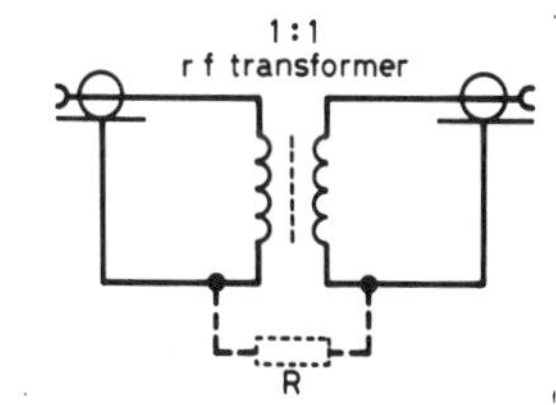

Another method of braid-breaking is a 1:1 ratio RF transformer as in Fig 9.4. This will only pass the wanted antiphase signal. One method of construction is to use two tightly coupled windings (usually twisted together) on a small ferrite ring or bead. Another is the 'Faraday loop' method, using two loops of coaxial cable taped together. However, great care must be taken to prevent excessive signal loss at higher frequencies. Home-constructed units will probably be unsuitable for UHF TV.

Fig 9.5. UHF TV braid-breaker using balanced high-pass filter technique. This circuit also attenuates interfering signals picked up on the antenna below 470MHz. L is 4 turns 22swg 5mm i/d. R is a static discharge path – 10kΩ or more

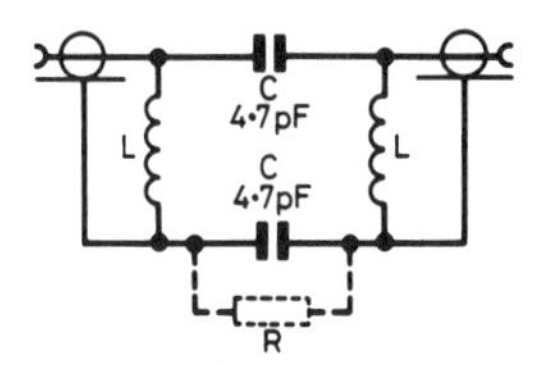

A third type of braid-breaker, primarily designed to protect UHF TV reception, is shown in Fig 9.5. Its operation is essentially different from the other. It consists of a balanced HPF with a cut-off of 470MHz, but performs satisfactorily in the unbalanced coaxial downlead. The low-value coupling capacitors have a high impedance at lower frequencies, and therefore block the unwanted RF currents.

Improving station design

Because the amateur station is likely to be in close proximity to domestic electronic equipment, it is in his interests to minimise the possibility of EMC problems.

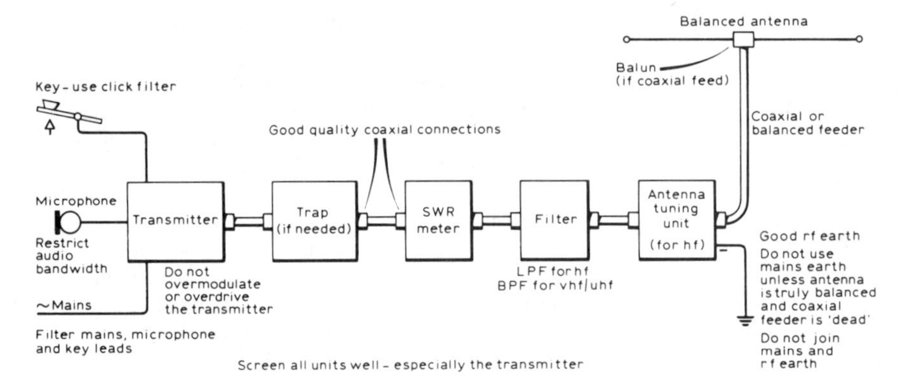

Fig 9.6. Example of station format required for good EMC

The amateur must obviously ensure that the levels of unwanted emissions from his station are acceptably low. However, even if the problems are due to inadequacies in the domestic equipment, it is desirable to try to design the station and antennas so as to keep the regions of high radiated field strength as far as possible from other domestic premises. While this is not always easy, it has a number of benefits. It minimises the chance of interference to domestic equipment, and of the station picking up noise from domestic equipment. It also improves the performance of the station – high field strengths in adjacent premises not only give rise to EMC problems, but also waste power that should really be going out to distant parts of the globe!

Particular attention should be paid to the type and location of the antenna, its feeder and how the station is earthed.

Fig 9.6 shows an amateur station designed for good EMC. Each part will be considered in turn.

Transmitter

Transmitter design which minimises harmonic and other unwanted outputs is dealt in Chapter 8. Because the higher harmonics usually fall off in level (see Fig 8.4), the UHF harmonics which are radiated by a well-designed HF transmitter are rarely responsible for television interference (TVI). Much greater attention must be paid to VHF and UHF amateur transmissions. For example, the fourth harmonic from the 144MHz band falls in channel 34 (575.25MHz vision, 581.25MHz sound) and the fifth in channels 52 (719.25/725.25MHz) and 53 (727.25/733.25MHz). However, unless the harmonics are particularly strong, this problem is likely to be apparent only in those areas where one of these channels is allocated.

The FM radio band (87.5 to 108MHz) may suffer from certain harmonics, particularly from the HF end of the 28MHz band (the third falls between 87.5 and 89.1MHz) and the new 50MHz band (where the second falls between 100 and 104MHz). There is also some possibility that strong third harmonics from the 3.5MHz band could affect the 10.7MHz IF stages in the radio.

Harmonics are not the only problem. Most transmitters use frequency multipliers or mixer circuits, and the output may contain certain signals having no harmonic relationship to the actual output signal (see Chapter 8). These too must be well suppressed.

Many transmitters claim, or in reality actually achieve, only 40–60dB of harmonic suppression, with other unwanted signals somewhat better. The ATU and antenna will provide some further attenuation. While this may be adequate in many circumstances, in some cases additional attenuation will be necessary. Usually this can be achieved by inserting an appropriate filter in the feed to the antenna system.

The effectiveness of additional filtering can be seriously impaired if unwanted signals are allowed to leak out of the transmitter. While the signal which is fed to the antenna may be 'clean', the signals which leak out may cause persistent, and apparently incurable, interference problems. Serious attention should be paid to the effectiveness of the screening inside the transmitter and of the case, and of the decoupling and filtering of the power, key, microphone and any other leads connected to the transmitter. These must not be allowed to carry RF energy from inside the case, and then allow it to be radiated.

Trap

A 'trap' is a simple stop or notch filter, often consisting of a single resonant circuit. It is therefore simple to make and to tune to the correct frequency. It is not always necessary, but it is an easy way of providing an additional 20 to 30dB of attenuation over a relatively narrow bandwidth – often very useful when a particular transmitter harmonic might cause a problem. At VHF and UHF, a tuned coaxial 'stub' (usually $\lambda/4$ long) is often used. Unless it is compensated, a trap may cause some mismatch. This is usually small, but to prevent a permanently high SWR reading, it is best to locate it before the SWR meter.

SWR / power meter

The SWR meter is described in Chapters 7 and 10 – in particular, see Fig 7.10 and associated text. With some designs, the sensitivity varies with frequency. With others, it is essentially constant over a wide frequency range, and this type is usually calibrated to measure RF power. Some meters will read peak envelope power (PEP) and are convenient for

SSB measurements. The SWR meter enables impedance match of the antenna and feeder system to be checked, or for the ATU (if used) to be adjusted for minimum SWR at its input, thereby presenting the filter with the correct load impedance at the transmitting frequency. The meter must be designed to operate with the same impedance as the filter (usually 50 or 75Ω). The positions of the SWR meter and filter could be interchanged but, as the meter uses diodes to rectify a sample of the RF signal passing through it, these may generate some low-level harmonics. It is therefore better to place it before the filter, which will then remove the harmonics.

At HF, where operation usually takes place on several bands, it is common practice to use a low-pass filter with a cut-off frequency of 30MHz. While this will not reject unwanted signals below 30MHz, it does protect the VHF radio band, and other services which might be affected even by low levels of unwanted signals. A few milliwatts at VHF can cause a surprising amount of interference over a wide area.

At VHF and UHF, although a few transmitters do cover more than one band, each band usually has its own antenna and feeder. While low-pass filters could be used, it is just as convenient to use a bandpass filter for each band. This will remove unwanted transmitter outputs either side of the amateur band. It will also help on reception where very strong radio transmissions from other services sometimes cause overload problems to sensitive amateur receivers.

Antenna tuning unit (ATU)

Strictly speaking, an ATU is a matching unit rather than a tuning unit, as explained in Chapter 7, it is also known as an 'antenna system tuning unit' (ASTU) or as a 'transmatch'. Opinions differ as to whether an ATU is necessary in situations where the antenna and feeder system presents the correct 50 or 75Ω impedance required by the transmitter, SWR meter and filter. However, the 30MHz low-pass filter in an HF station will not remove those transmitter harmonics which fall inside its passband, eg when operating on 14MHz and below. It is unlikely that the harmonic suppression of the transmitter alone will exceed 50dB. Depending on the type of circuit used, an ATU can give some useful additional attenuation of these harmonics and of other unwanted outputs.

If the ATU circuit is such that the output is inductively coupled, the antenna and feeder system can (if required) be isolated from the rest of the station equipment. If twin feeder is used, the ATU performs the function of a balun (see later). If coaxial cable is used, the screening braid can be earthed either to the equipment earth or to a separate RF earth, as required. If the ATU feeds directly into an end-fed wire (note that this is not recommended) it enables the RF earth return to be kept quite separate from the mains earth. Fig 9.7 shows an example of this situation.

Feeder, antenna, balun and earth return

While it is the RF current flowing in the antenna which is responsible for true electromagnetic waves, those parts of the antenna which are at a high RF voltage will produce a strong, localised voltage field. This can transfer the transmitted signals into nearby equipment and wiring because of the effects of capacitive coupling. Similarly, those parts carrying high currents may cause considerable inductive coupling. To minimise the problems that may result, the antenna should be situated as far away as possible from buildings and from any overhead mains and telephone wiring running outside, even if this means using a somewhat longer run of feeder. A relatively small increase of distance can result in a considerable reduction in the level of pick-up of the amateur signal (see Chapter 7).

Where a high-gain beam antenna is used (particularly at VHF and UHF), avoid situations where it will 'fire' straight into a house and the electronic equipment inside it. Mount the antenna as high as local planning regulations allow, so that it beams over the rooftops and, if possible, over neighbouring receiving antennas. Note that is often difficult to cure problems which are due to direct pick-up in the affected equipment, but relatively easy if the pick-up is on the antenna itself.

Antennas most likely to increase breakthrough are those which are indoors, cross over the house, or are end-fed directly from the house. These are not recommended. If the transmitting equipment is above ground level (the spare bedroom is a popular location), do not forget that the RF earth return for an end-fed wire actually forms part of the antenna system, and will radiate strongly. Fig 9.8 shows a typical domestic situation where an end-fed wire is used. This may be satisfactory for receiving, but could cause problems when transmitting.

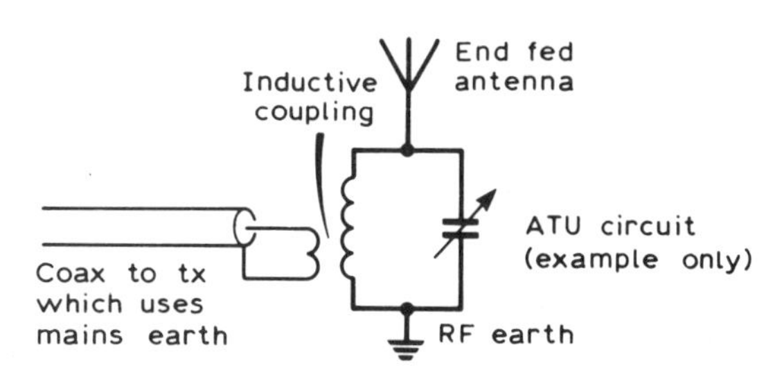

Fig 9.7. Use of inductively coupled ATU circuit to enable the RF and mains earths to be separated. Note that the RF earth is part of the antenna system and carries all the antenna current

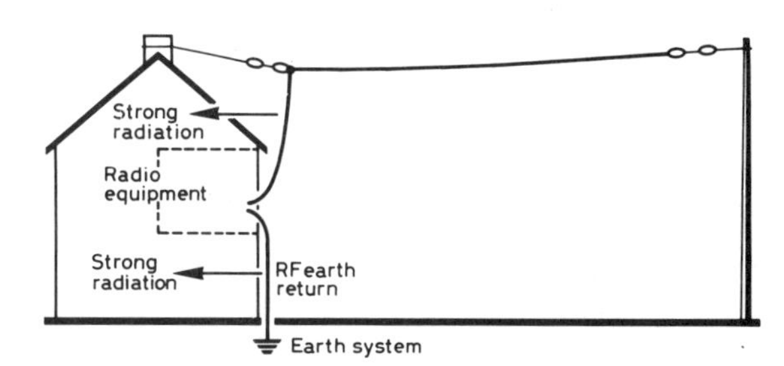

Fig 9.8. Typical end-fed wire antenna (often used by shortwave listeners). This results in strong radiation into the house and can cause problems when transmitting

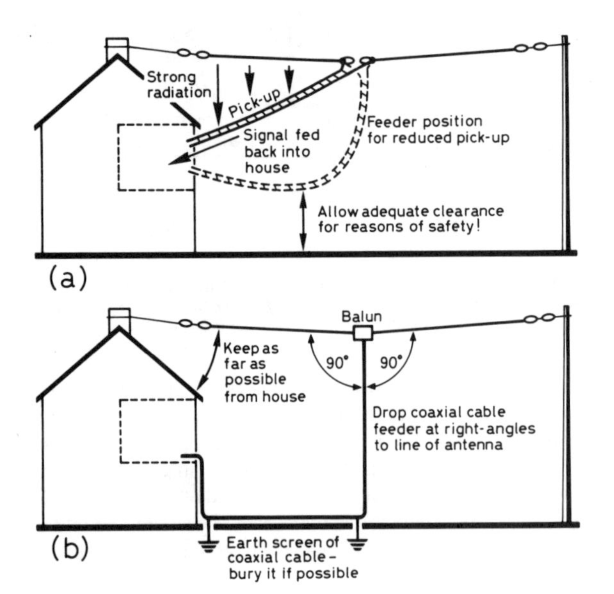

Fig 9.9. Prevention of pick-up of the radiated signal by the feeder: (a) by maximising the distance between antenna and feeder; (b) by earthing the coaxial screen

A horizontal, balanced antenna (eg a λ/2 dipole) will usually produce a lower field strength around the house than an end-fed wire. It should be fed with balanced (twin) feeder or, if the feed impedance is suitable, with coaxial cable via a balun (balance-to-unbalance transformer) at the centre of the antenna (see Chapter 7). Although a dipole will work without a balun, a considerable amount of RF current will flow back from the antenna on the screening braid of the cable, causing it to radiate strongly. To make matters worse, the cable is usually connected through the station equipment to the mains safety earth. The braid current will therefore flow into the mains wiring, and re-radiate the signal throughout the house.

To prevent the signal which is radiated by the antenna from being picked up by the feeder and fed back to the house, the feeder should drop at right angles away from the antenna for as far as possible. It should not be pulled tight so that it runs close to, and under, one 'leg' of the dipole, as in Fig 9.9(a). Coaxial cable (and also screened twin, which is rarely used by amateurs) has an advantage over unscreened feeder because it can lie in contact with the ground, be buried and (if possible) earthed, as shown in Fig 9.9(b). This ensures that no current is fed back.

If it is essential to use an end-fed antenna, do ensure that the RF earth and the domestic mains earth are kept separate. Operating at ground level from a remote outbuilding (such as a garden shed) allows short, direct earth connections to be made. An alternative is to use an earthed coaxial cable to feed the antenna, as shown in Fig 9.10, although matching at the antenna feed point (especially if several bands are used) does require some ingenuity. A remote tuning unit is one solution. Another is to use a trap monopole antenna, ie one half of the trap dipole in Fig 7.13, or a variant of it.

If the problems still persist after taking steps to optimise the design of the station, it may helpful to come to some acceptable agreement to operate only at selected times or at power levels where the problems do not occur, while efforts are made to diagnose and cure the problem at the affected equipment.

EMC problems tend to appear or vanish with relatively small changes in power level, and it may be that a slight reduction of power will avoid most of the problems. Problems can also be minimised by being more flexible in choosing power levels. It is not necessary to use as much power for cross-town contacts as for working long distances.

For this reason, a speech processor can be of great help, particularly on SSB. It increases the average value of the speech waveform while keeping the peaks to a predetermined maximum. This produces a more 'punchy' signal which can provide some of the benefits of higher output power without the EMC problems.

The affected equipment

If interference persists after all reasonable efforts have been taken to maximise the EMC of the station, then attention must be paid to the affected equipment and its installation conditions.

How and where interference occurs

When the amateur signal is adequately free of all unwanted outputs, any problems which occur are usually due to the inability of the equipment affected to operate in the RF fields generated by the transmitter. It is perhaps not too surprising that, unless the manufacturer has provided the necessary safeguards, circuits which are designed to work at millivolt or milliamp levels (or even less) are adversely affected when signals maybe 1000 times stronger are present. However, in today's dense electromagnetic environment, such safeguards should be included in all electronic equipment. Interference falls mainly into the following categories:

(a) *Direct reproduction.* The frequency of some amateur signals is within the passband of certain stages. If the amateur signal enters one of these stages it is simply added

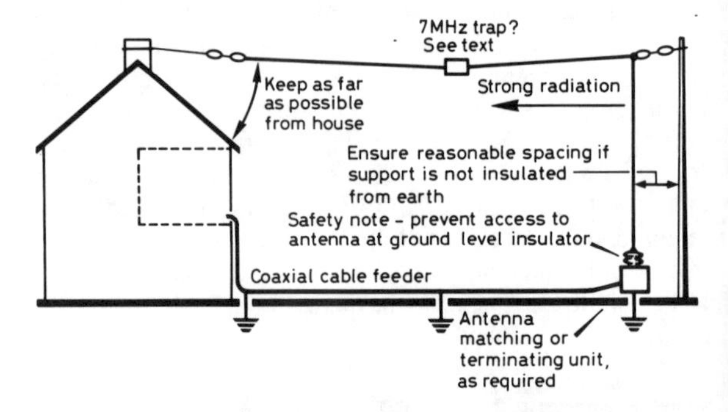

Fig 9.10. End-feeding an antenna while maximising EMC. Although less convenient than Fig 9.8, this method causes far fewer interference problems

to the wanted signal. For example, the 1.8, 3.5 and even the 7MHz band may affect the video frequency circuits in TV sets and video recorders (often causing finely spaced diagonal bars on the picture). A 3.5MHz transmission may enter the colour circuits centred on 4.43MHz (the picture sometimes reverts to black and white when the receiver is affected). The 10MHz band is close to the FM radio IF on nominally 10.7MHz. Note that in many cases a relatively low level of unwanted signal flowing in the circuit may produce a significant amount of interference.

(b) *Harmonic generation*. This normally requires a somewhat higher level of unwanted signal to be present in the affected stages, ie sufficient to over-drive the circuit so that it becomes non-linear. Harmonics can be generated in RF and IF amplifiers, mixers etc (particularly in TV and radio tuner stages). The effect is similar to that when the amateur signal itself contains harmonics.

(c) *Cross-modulation, desensitising and blocking*. Again this is an effect of relatively strong signal pick-up and subsequent non-linearities. The presence of the unwanted signal causes the gain of the affected stage to change (usually it is reduced). Variations of the level of the unwanted signal – such as occur when transmitting SSB, AM or even CW (A1A mode) – are therefore impressed on the affected signal. This can be very disruptive. This effect is less noticeable where the transmission amplitude is constant (eg FM).

(d) *Rectification (detection, demodulation etc)*. As with (b) and (c), non-linearities are responsible. The problem is most noticeable in audio circuits where the base-emitter junction of a bipolar transistor (either as a discrete device or as part of an integrated circuit) acts as a rectifier when strong RF signals are applied to it. Again, SSB, AM etc are more troublesome than FM.

(e) *Timing errors*. Sometimes it is the timing and sequence of signals which enables a circuit to operate correctly. In this category are digital circuits, TV timebases etc. Usually the signals are pulses of several volts in amplitude, so the circuits are relatively immune to RF breakthrough provided that this is below a certain level. However, above this level the timing may be upset by the RF signal superimposed on the pulse waveform, and this will cause the circuit to perform incorrectly. For example, the timebase circuit may fail to trigger correctly, causing the picture to 'tear' or roll. Circuits involving microprocessors and computers may fail to carry out the expected control functions etc.

Interference pick-up and paths

The three main routes are direct pick-up, pick-up via connecting leads, and indirect pick-up. Take, for example, a tape recorder which, while running on batteries alone (with no mains lead), suffers breakthrough. This is probably a case of direct pick-up in the circuitry. Apart from moving it around the house to try to find a place where the field strength of the amateur transmission is lower, or carrying out modifications to the recorder itself, there is probably little that can be done to cure the problem.

If the breakthrough only occurs when the mains lead is plugged into the recorder, and ceases when it is disconnected, this is generally a good sign. It means that the problem can probably be cured by fitting a filter or ferrite ring choke in the lead.

Indirect pick-up is more difficult to diagnose and to cure. The recorder may be satisfactory when running from batteries, but the interference starts even when the lead is brought near it, before it is actually plugged into it. It is obvious that fitting a filter in the lead will probably have little or no effect. One method of dealing with this problem is dealt with later.

On the lower-frequency bands, the unwanted signal is most likely to be picked up and fed to the affected equipment via the leads connected to it. The cause of the interference may be the RF current flowing through the circuit or, less commonly, the increased voltage field around the equipment. Both may occur at the same time. Longer leads generally mean more pick-up. Even when the RF current passes only through the 'earthy' side of the circuit, this can create voltage differences within the circuit.

At the higher frequencies (particularly VHF and UHF), because parts of the circuit and the internal wiring are long enough to act as efficient receiving antennas, there is more danger of direct pick-up in the equipment itself.

Interference to television (TVI)

In the UK, TV is broadcast only at UHF (470 to 860MHz). The TV antenna is too short to be an efficient 'receptor' on the lower-frequency amateur bands (ie below 30MHz). It is much more likely that pick-up occurs on the braid of the TV antenna downlead, which acts rather like an end-fed receiving antenna (typically about 30ft long). Strong RF currents enter via the 'outer' side of the TV antenna coaxial socket, flow through the set and into the mains wiring. Almost any part of the circuit may be affected. A conventional high-pass filter, fitted in the coaxial socket at the TV set antenna input, filters only those signals which are picked up on the antenna itself, so it will probably be ineffective. However, one of the braid-breaker circuits in Figs 9.3, 9.4 or 9.5 is much more likely to be successful. Occasionally a ring choke may be needed on the mains lead.

At frequencies above about 30MHz, the TV antenna becomes an increasingly efficient receptor of unwanted signals (especially on the 144 and 432MHz bands), while braid pick-up problems tend to decrease. The tuner stage in the TV set may be overloaded. Self-generated harmonics usually affect only certain TV channels just as though the amateur transmitter was at fault (see examples given previously). However, in cases of cross-modulation (and where interference occurs after the tuner), it is usual for all the channels to be affected more or less equally.

If it is the TV antenna which is picking up the unwanted transmission, a filter fitted at the antenna socket of the set should always be effective, provided that it has sufficient rejection on the unwanted frequency. For multi-band operation up to the 432MHz band the best choice is probably a good-quality UHF high-pass filter which has a sharp cut-off below 470MHz. However, if interest is concentrated on a particular band, a bandstop filter could be used. Fig 9.11 shows a simple circuit which is easy to construct and which can be aligned without test equipment.

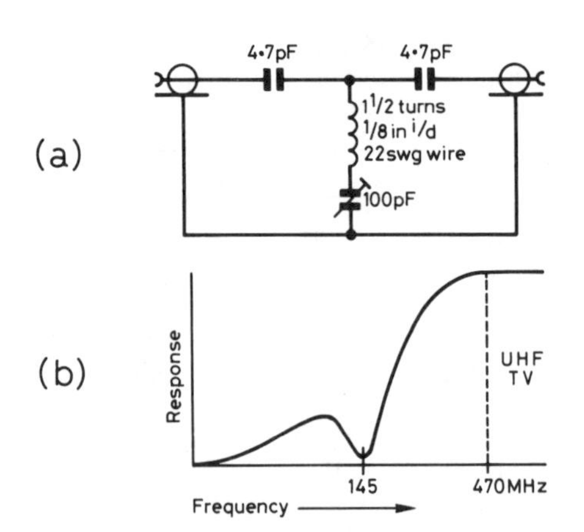

Fig 9.11. A high-pass TV filter with increased rejection of a particular band of interest – in this case 144MHz

Difficult cases

While some cases of interference are readily cleared by the fitting of a filter or a braid-breaker (sometimes both are needed), others can be more difficult. Braid-breakers and ring chokes will only block the interfering currents. Occasionally it is the RF voltage field introduced into the vicinity of the set – and not the current – which is responsible for the interference (ie an indirect pick-up effect). Indeed, blocking the current may actually increase the voltage – and the breakthrough. In such cases it may be better to fit the braid-breaker a short distance along the coaxial cable, away from the set (but not too far – 10ft maximum). Similar considerations apply to the positioning of a ring filter on the mains lead. All other non-essential conductors which may also be bringing interference to the set (eg a set-top lamp) should also be removed. If it is electrically safe to do so, connecting the braid of the down-lead to a reasonable RF earth will often help greatly by diverting the interfering signal away from the TV set.

Broadcast radio interference ('BCI')

Transmissions on almost any amateur band may break through on domestic radios. A classic case is 'image' or 'second-channel' interference (see Chapter 5) to medium-wave reception from 160m transmissions. For example, when trying to listen to 1053kHz (one of the frequencies currently allocated to BBC Radio 1), the following occurs:

Wanted mixer product

Local oscillator – Wanted frequency = IF
1508kHz – 1053kHz = 455kHz

Unwanted mixer product

160m signal – Local oscillator = IF
1963kHz – 1508kHz = 455kHz

Even when operating on the higher frequency amateur bands, with some receivers there is still the possibility of interference to medium-wave reception (rarely to the long-wave). The receiver local oscillator often contains so many harmonics (in any case these will be produced in the mixer stage) and each time that one falls 455kHz away from an amateur signal, a 455kHz IF is produced.

A satisfactory cure would be to reject the amateur signal before it reaches the mixer, but in practice this is not normally possible. Most sets have no external antenna into which a filter (eg 1600kHz low-pass) could be inserted, and internal modifications would be a formidable task. However, it may be possible to reduce the interference by treating it as a case of direct pick-up, and keeping the radio away from wiring, piping (which may be increasing the strength of the interfering signal), by experimenting with its orientation etc. As the problem only occurs on certain 'spot' amateur frequencies, it may be no great inconvenience to avoid these at times when it is known that annoyance could occur.

VHF-FM problems may usually be treated by the same techniques as used for interference to television, provided that a properly installed Band 2 antenna is used. The band is from 87.5 to 108MHz. However, as the antenna is longer, it will pick up more HF amateur signal than a TV antenna. In equipment where the radio facility is part of a stereo hi-fi system, the possibility of RF breakthrough directly into the audio circuit must not be overlooked. This is dealt with later.

Self-generated harmonics which are troublesome are the third from the HF end of the 28MHz band (affecting 87.5 to 89.1MHz) and the second from the 50MHz band (affecting 100 to 104MHz). Usually only one programme is affected at any one time, depending on the amateur frequency in use.

Interference from the VHF amateur bands (where considerable antenna pick-up is likely) is also a problem. The major cause will be cross-modulation, and usually the whole band is affected. Braid pick-up may be cured by the braid-breaker circuits of Figs 9.3 or 9.4. For antenna pick-up, a suitable filter may be selected from the spectrum chart of Fig 9.1. To reject HF transmissions, a high-pass filter with an f_c of 87MHz would be ideal, but one with an f_c of 40MHz (originally intended for use with VHF TV sets) should also be effective. To reject a particular VHF or UHF amateur band, a bandstop filter could be used. However, a more universal solution would be to use a good bandpass filter which covers the whole of the FM radio band.

Some FM radios have antenna inputs designed for 300Ω twin feeder. In many cases perfectly satisfactory reception is obtained if 75Ω coaxial cable is connected directly to such sets but, when additional filters are needed, it may also be beneficial to fit a 75 to 300Ω balun.

Audio equipment

Audio hi-fi equipment often suffers from RF breakthrough, whether it incorporates a radio tuner or not. Although the screening, filtering, decoupling, earth paths etc in the equipment may be adequate for audio purposes, they are generally poor at RF. Sensitive audio amplifiers are easily driven into non-linearity by high level RF signals, and rectification occurs. With AM (A3E) transmission, the hi-fi owner will probably hear every word transmitted, and possibly at considerable volume. If it is SSB, only the characteristic 'quack quack' will be heard. With FM, there may only be a click when the transmission starts and finishes, and this may often go unnoticed. However, in cases of serious overload, the audio volume may decrease, or even be totally silenced.

Where the audio equipment consists of separate units, it is usually possible to filter the interconnecting leads, but when it is one integrated system, such possibilities are limited. Where the leads are accessible, using the ferrite ring choke technique to block longitudinal unwanted RF currents is often effective.

In stereo systems, the loudspeaker leads are frequently the cause of the trouble. They form a dipole antenna which is only too efficient at picking up radio transmissions, and may even resonate on some amateur bands. Strong RF currents are fed through the audio amplifier, and these affect particularly the negative feedback circuits. A ring choke inserted in each lead close to the amplifier speaker output socket is often very successful. It may also be useful to fit a decoupling capacitor of 1000 to 10,000pF (1 to 10nF) across each lead at the connecting plug. In difficult cases, screened speaker leads may be necessary.

Sometimes conventional filtering may be necessary to remove the RF interference picked up, for example, by a magnetic pick-up in a record turntable unit. As this has a low level AF output (a few millivolts) and requires considerable amplification, it is rather susceptible to such problems, especially from VHF transmissions. Fig 9.12 shows a suitable circuit.

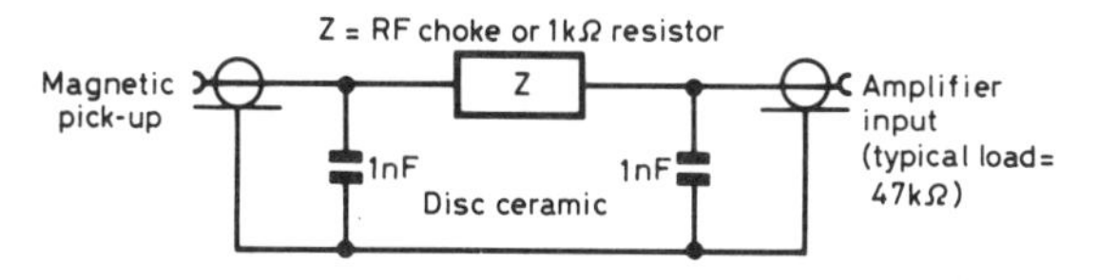

Fig 9.12. A low-pass filter to remove pick-up on a magnetic cartridge

Although internal modification to any equipment is the responsibility of its owner and not the amateur, this may be necessary in stubborn cases of interference where internal pick-up is diagnosed. Fig 9.13 shows examples of RF filtering directly on the affected device.

Video cassette recorders (VCRs)

These have problems in addition to those suffered by TV sets. Most recorders incorporate a wideband antenna preamplifier which covers all UHF – and possibly VHF – TV frequencies. This may be susceptible to overloading by strong amateur signals (especially VHF and UHF). On replay, the high-gain video circuits which amplify the weak signals delivered by

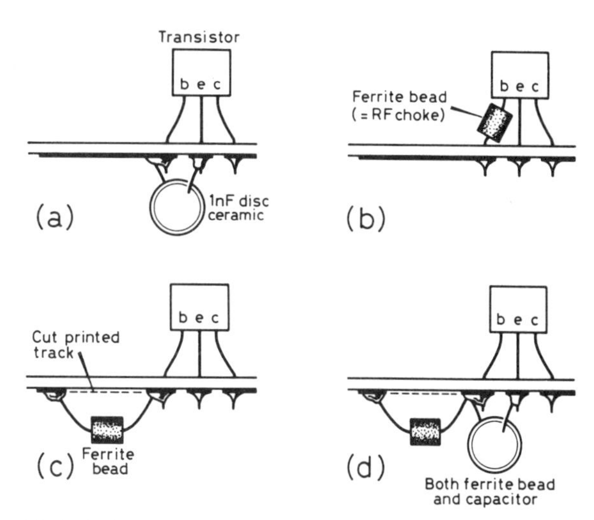

Fig 9.13. Examples of internal filtering sometimes needed to cure stubborn cases of interference – especially from VHF and UHF transmissions. High-gain, low-level audio stages are particularly vulnerable

the heads may suffer from the lower frequency bands – particularly 1.8, 3.5 and 7MHz. Note that filters and braid-breakers may have to be fitted both to the antenna lead and the lead to the TV set.

Masthead preamplifiers

These are sometimes used in areas of poor reception of TV and FM radio. They are normally fitted directly at the antenna, and powered via the coaxial downlead. Many cover the entire VHF and UHF spectrum, and some respond down to a few megahertz. Attempts to cure interference problems at the TV or radio will not be fully successful if interference which occurs because of amplifier overload is not dealt with beforehand. Normally, fitting a suitable conventional filter at the preamplifier input is effective but, as the antenna will probably be mounted in a well-elevated position, the practical problems of doing this are obvious.

Home distribution amplifiers

It is becoming increasingly popular to amplify TV and radio signals, and then split them to feed several rooms in the house. As with masthead units, most systems are essentially wideband and are easily overloaded. The same treatment applies if access to the amplifier is possible. However, the use of multiple downleads may aggravate the effects of braid pick-up.

Other equipment

Interference to and from other equipment is a growing problem. The use of home computers and word processors is well established and interference problems with radio amateurs do occur, both in the form of breakthrough into the computer, and of signals radiated by the computer which can interfere with the amateur's reception. Troubles may be minimised by fitting ring chokes (or some form thereof) to the mains leads, and especially to the leads interconnecting the peripheral units.

Social aspects of EMC problems

Test transmissions

On obtaining a licence the amateur is extremely unwise to start making transmissions at the maximum permitted power level without having due regard to the amount of interference which might be occurring. He should not adopt the attitude that he is licensed to run full legal power and will do so, regardless of the consequences. His first concern should be to find out the extent of his problems, and the best place to start is his own home.

The amateur licence states that interference tests should be carried out 'from time to time'. It is a good idea to do tests particularly when changes are made to the station equipment or antenna system, or when different bands are operated. This obviously includes putting the station on the air for the first time. In order to minimise possible annoyance, initial tests should be at low power and of short duration, preferably at times when neighbours are unlikely to be affected. If relationships are favourable, the amateur may feel it is appropriate to inform them of these tests so that he can be told without delay of any signs of interference. On the other hand, he should beware of causing unnecessary alarm to someone who does not know him very well, about a problem which may be less severe than anticipated. It may be helpful to remember to refer to the problem as one of 'breakthrough' (which suggests an unintentional happening) rather than 'interference' (which may be understood as a deliberate act). The licence conditions state that the periods of test transmissions should be logged; 'EMC test' is a suitable entry. It is useful to record in some detail the test conditions, such as the transmitter and other equipment used, power output, antenna system, interference observed, effectiveness of filtering, etc. If lengthy, these records could be entered as a dated, separate sheet in the back of the log book. If the station is ever checked for interference by the DTI, they will show that the amateur has tried to avoid 'undue interference' as required by the licence.

Convincing your neighbours

If the amateur is able to operate his station without causing interference in his own home it will help to convince others that his activities do not inevitably lead to trouble. In the early days of his operation he may not be able to run full power on all the bands available to him and some of his domestic equipment may still be affected, but his household should be able to use at least the basics, such as the family TV set, radio and audio system, when he is on the air.

Meeting your neighbours

If a neighbour has started to experience interference, then his reaction will probably be reasonably friendly provided that the transmissions have been for only short periods. However, he is unlikely to know anything about interference or breakthrough problems and will expect the amateur to 'put a stop to it'. It is at this point especially that the amateur is advised to conduct himself with the greatest diplomacy and tact. Good neighbourly relationships can suddenly turn sour and develop into protracted confrontation. This must be avoided at all costs. The amateur should try to avoid making too much of an apology for the interference, or any long-term commitment to avoid transmitting at certain times as both may be interpreted as an admission of blame. If he is sure that his station is being operated with due regard to the problems of EMC, in principle the amateur could simply refer the complainant to the DTI Radio Investigation Service. However, as this could result in unnecessary expense for the complainant and some interruption of the amateur's transmitting activities, it would be far more neighbourly to offer to co-operate with the neighbour to try and solve the problem between themselves in the first instance.

If the offer of assistance is accepted, the amateur will probably have to make several visits to the home of the complainant with a view to fitting filters etc to the affected equipment. Such visits should not be protracted, but carried out in an efficient, business-like manner. The assistance of another radio amateur (preferably one who recognises the symptoms of interference) helps greatly in the diagnosis of the problem. If possible, a selection of filters and other devices should be taken (kits are often owned by radio clubs for the purposes of diagnosis).

To reduce the likelihood of recurring complaints, the opportunity may be taken to fit filtering for all the bands which the amateur expects to use. For example, if the present complaint is TVI due to overload from 144MHz transmissions, it would be wise to fit a UHF high-pass filter which is known to be effective up to the 432MHz band, and possibly also a braid-breaker in case interests turn to the HF bands. Test transmissions, conditions and results (successful or otherwise) should again be logged.

It must be stressed that the amateur should fit filters and make any adjustments, modifications etc to the affected equipment (even on a temporary basis) only if the complainant is perfectly happy for him to do so. He may still be rather unsure about the situation and about the amateur's technical ability. On his part, the amateur should be aware that unless he is careful, he could be thought responsible for subsequent faults which occur in the equipment which he investigates. He is strongly advised not to make any internal modifications to any equipment, even if he is technically competent and qualified to do so.

Having demonstrated that a filter provides a satisfactory cure, the amateur is not obliged to provide one free to the complainant. However, as the cost is probably minimal, in the circumstances he would be wise to consider himself fortunate, and not even think about asking for payment for the filters.

If it appears that the problem cannot be dealt with amicably, or if faced with technical problems, the amateur would be wise to contact the RSGB EMC Committee which can provide advice in such cases. They have accumulated considerable experience in dealing with domestic EMC problems, and may know of proven solutions for some types of equipment. Local clubs may also be able to provide assistance.

DTI assistance

The Department of Trade and Industry have produced the booklet entitled *How to improve television and radio reception*, parts of which are applicable to interference from radio amateurs. It is available from main post offices. It is recommended that the amateur obtain a copy. The DTI Radio Investigation Service (RIS) may be requested to deal with complaints of amateur interference, although in practice they become involved in only a minority of cases, such as:

(a) when the complainant does not know the source of the interference;
(b) when the complainant refuses help, or does not tell the amateur about the problem;
(c) when the amateur is unable to provide a satisfactory cure.

This service requires payment of a fee by the person requesting assistance. Officially it covers only TV and radio reception. Details are given in the booklet. If the complainant identifies the amateur station as the source of interference, the RIS will check the station without charge to either party. However, if the station is 'cleared' and the complainant wants his equipment to be checked, he will probably have to pay a fee for the diagnosis of the problem, and even then the RIS may be unable to provide a cure. This is why he is well advised to co-operate with the amateur. On the other hand,if it is obvious that relationships are or will become hostile, the amateur should avoid unnecessary antagonism and encourage the complainant to refer the matter to the RIS.

Safety precautions with mains supplies using protective multiple earthing

Throughout this chapter and Chapter 8, various recommendations are made concerning earthing both at the transmitting station and at the affected equipment. These are intended to minimise interference problems, and such techniques have been used for many years with considerable success. However, some local electricity supplies now use the system known as protective multiple earthing (PME), and following these recommendations may contravene the safety regulations for such systems. There is a serious risk of injury or death occurring under certain fault conditions. Accordingly, before any earth other than or in addition to the normal mains earth is connected to any equipment or system it is extremely important to check that PME is not being used. If it is, then further advice on how to achieve the EMC objective must be sought.

CHAPTER 10

Measurements

Correct operation of amateur radio equipment involves measurements to ensure optimum performance, to comply with the terms of the amateur transmitting licence and to avoid interference to other users. The purpose of these measurements is to give the operator information regarding the conditions under which his equipment is functioning. There are three basic parameters: voltage, current and frequency.

For example, in even the simplest transmitter it is necessary to know the drive to the various stages (current measurements), the input power to the PA (current and voltage measurements) and the frequency of the radiated signal.

DC measurements

The basic instrument used for the measurement of voltage and current is the moving-coil meter. This comprises a coil of wire, generally wound on a rectangular former, which is mounted on pivots in the field of a permanent magnet (Fig 10.1). The coil develops a torque proportional to:

(a) the current flowing through it;
(b) the strength of the field of the permanent magnet.

Current is fed to the coil through two hair springs mounted near to each pivot. These springs also serve to return the pointer to the zero position (on the left-hand side of its travel in standard meters) when the current ceases to flow. Provision for adjusting the position of the pointer is made by a 'zero adjuster', accessible from the front of the instrument.

It is usual to 'damp' the coil system (ie prevent it swinging freely after a change of current), a common method being to wind the coil on an aluminium former, which then acts as a short-circuited, single-turn coil in which the eddy currents serve to oppose the movement. The degree of eddy current damping is also dependent on the external resistance across the terminals of the moving coil and is greatest when the resistance is low. It is a wise precaution to protect sensitive instruments not in use by short-circuiting the terminals.

The scale of the moving-coil instrument is linear; it can only be used on DC, but can be adapted to measure AC.

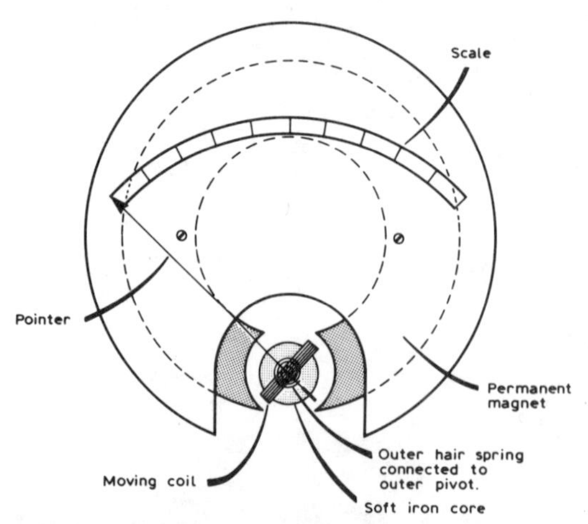

Fig 10.1. Construction of moving-coil meter

Milliammeters

Milliammeters and microammeters are commonly manufactured with full-scale deflections (FSDs) between about 25μA and 10mA. For higher current ranges, a shunt resistor is connected across the meter (see Fig 10.2(a)).

Voltmeters

A milliammeter may be used to read DC voltages by connecting a resistor, termed a 'multiplier', in series with it (see Fig 10.2(b)).

The accuracy of a meter depends on many factors, eg size, quality, accuracy of shunt or multiplier resistor, the scale deflection (ie full-scale reading). Usually, the size of meter most likely to be used in amateur radio would be 60–75mm in diameter and the accuracy would be of the order of 3%.

If the resistance of a voltmeter is too low, it is likely that it will pass more current than the circuit whose supply voltage is being measured, the voltage measured at that point will

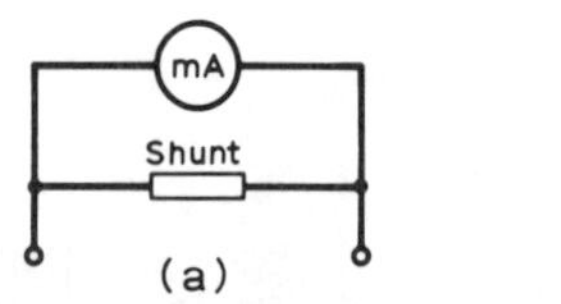

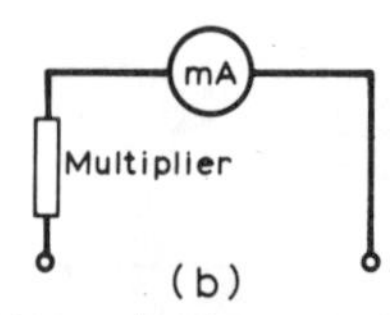

Fig 10.2. Extending range of an MC meter (a) to read higher current with parallel shunt, (b) to measure voltage with series resistor or multiplier

therefore be lower than it actually is. Thus the sensitivity of a voltmeter is usually defined in 'ohms per volt' (Ω/V). For example if a voltmeter has a resistance of 20,000Ω (ie meter and multiplier) and a full-scale reading of 100V, its sensitivity is 20,000Ω/100V or 200Ω/V. This value is much too low for use on radio circuits. A satisfactory value is 20,000Ω/V which corresponds to a basic meter full-scale deflection of 50μA.

AC measurements

Appreciable errors arise if the waveform of alternating voltages or currents to be measured is not sinusoidal. For example, a meter which measures the peak value of a voltage is likely to be calibrated in RMS values; if the waveform is not sinusoidal, the ratio between the peak and RMS voltages departs from the sine wave value and the calibration is in error by an amount which depends on the degree of distortion in the waveform.

The moving-coil meter can be adapted to measure AC by the addition of a bridge rectifier known as an 'instrument rectifier'. Although such a meter indicates the average value, ie 0.636 of the peak value of a sine wave, it is calibrated in RMS values. Rectifier meters may be used at frequencies up to about 10kHz, but are only accurate when the waveform is sinusoidal. Multiplier resistors can be added to a rectifier instrument but current measurement requires a special transformer known as a 'current transformer'.

Thermocouple meters

The thermocouple is a junction of two dissimilar metals which when heated generates a DC voltage. The junction is heated by the current to be measured passing through a 'heater' to which it is attached. In conjunction with a thermocouple, a moving-coil instrument can be used to read alternating currents of up to radio frequencies. A disadvantage is that low current readings are rather severely compressed. Thermocouple instruments read RMS values irrespective of waveform. Unless specially designed, they become less accurate as frequency increases, owing to the effect of the shunt capacitance.

These meters must be used with great care as the thermocouple itself can be burnt out by a current not much greater than the maximum reading of the meter.

Measuring and test equipment is now available with digital read-out at low prices. The basic accuracy of these depends upon the standard of the internal electronic circuits used but the overall accuracy is very high because the read-out is much easier to read accurately than a pointer instrument (or mechanical dial).

A disadvantage is that it is not easy to adjust to a maximum or minimum voltage/current using a digital instrument.

Multi-range meters

Many commercial multi-range meters are available. These consist of a number of voltage multipliers and current shunts, an instrument rectifier and current transformer, all of which are switched to provide a large number of DC and AC voltage and current ranges on a high-sensitivity meter. Digital multi-range meters are also available.

Voltage measurement at high frequencies

The rectifier instrument is usable at frequencies in the lower audio range and has a reasonable accuracy (order of 4%) provided that the waveform to be measured is sinusoidal. In the RF range an electronic voltmeter is required. This consists of a diode detector which produces a direct voltage proportional to the peak value of the alternating voltage. This is followed by amplifier circuits feeding a meter calibrated in volts. The electronic voltmeter can give accurate readings up to a frequency of several hundreds of megahertz and may have a sensitivity of the order of 10MΩ/V. The load it places on the circuit is therefore so small that it can be ignored, but the capacitance of the diode circuit becomes important at very high frequencies.

Measurement of voltage and current in the transmitter

The tuning or setting up of a transmitter is the resonating of the various tuned circuits in the transmitter at the required frequencies.

The low-power, fixed-frequency oscillators, mixers or frequency multiplier circuits are often pre-set, particularly in commercially built equipment when they are set up during manufacture.

Low-power semiconductor stages may be resonated by tuning the input for maximum collector current and then resonating the collector tuned circuit for minimum collector current. Thus facilities for measuring collector current must be provided. This is most conveniently arranged by a 'test point' in each stage to which a current meter of the appropriate range may be switched or connected.

A meter of a suitable range can be wired permanently into the PA collector circuit. This circuit may require to be retuned whenever the frequency band is changed. The PA current must be checked to ensure that the DC input power does not exceed the required value, for of course DC input equals voltage times current. It is satisfactory to measure the PA collector voltage at the DC input terminal of the PA, otherwise the application of a voltmeter to the collector connection itself will throw the circuit off-tune and give an erroneous reading.

Dummy loads

A dummy load consists of a non-inductive resistor having a dissipation equal to the expected output power. Such a dummy load would have a resistance of about 50 or 70Ω and may consist of a parallel or series/parallel connection of carbon resistors; for example, ten 5W 680Ω resistors may be connected in parallel to provide a 50W 68Ω load which would be satisfactory for a 75W input transmitter. The dissipation of the load may be increased by immersing it in oil. A dummy load should be screened.

Frequency measurement

The reasons for measuring frequency in the amateur station are:

(a) to ensure that the tuned circuits in the receiver or transmitter cover the required frequency range;
(b) to ensure that the frequency calibration of the receiver is correct;
(c) to ensure that transmissions from the station are within the licensed bands.

Assuming that a receiver having a reasonably accurate frequency calibration is available, the calibration can be checked against shortwave broadcast stations of known frequency or against standard frequency transmissions, for example MSF Rugby (2.5, 5 and 10MHz) or WWV (10, 15, 20 and 25MHz).

An accurate calibration may also be made by listening to the 'pips' produced in the receiver by harmonics of a crystal oscillator. When used in conjunction with a general-coverage receiver, a 100kHz crystal is usually adequate for checking frequencies up to about 4MHz. For higher frequencies, the spacing between 100kHz marker points may be too small to resolve, and a crystal of 500kHz or preferably 1MHz should be used in addition. If the receiver covers only the amateur bands, the bandspread is normally adequate to resolve the harmonics from the 100kHz crystal.

100kHz crystals for use in frequency standards are available with frequency accuracies of 0.002%. One of these should be considered as the prime frequency standard.

A calibration graph or table can now be drawn up for each range of the receiver. It must be kept in mind that the accuracy of this depends on the precision of the receiver dial mechanism, its logging arrangement and the presence of bandspread. The frequency of operation of a transmitter may then be measured using the receiver, calibrated as above.

Obviously the receiver cannot be tuned to even a low-power transmitter in the immediate vicinity, but it can 'listen to' the output of the VFO, the frequency of which may then be determined and hence the final output frequency of the transmitter. Although the VFO is screened, it will be found necessary to considerably reduce the receiver RF gain. The tuning range of a VFO can be quite easily adjusted to the required value in this manner.

The absorption frequency meter

The absorption frequency meter consists of a coil tuned by a variable capacitor, with a scale calibrated in frequency. It operates by absorbing power from a tuned circuit when tuned to the same frequency as that circuit. The tuned circuit must therefore be activated, and the absorption of power is indicated by a small change or flick in the transistor current flowing through the tuned circuit as the resonance point is passed.

Such a frequency meter has the advantages of:

(a) rugged construction and simplicity;
(b) low cost;
(c) no power supply required;

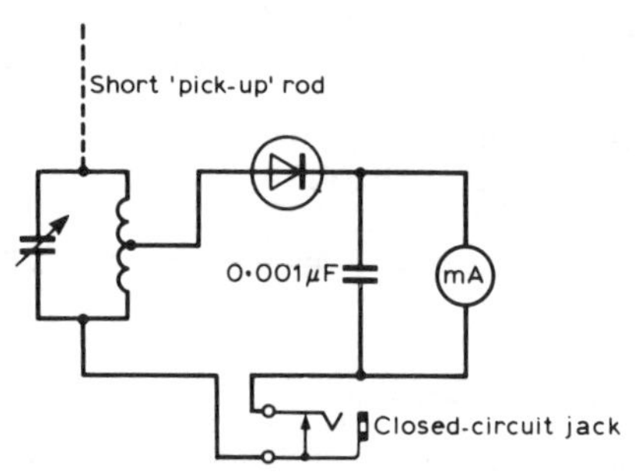

Fig 10.3. Absorption wavemeter with microammeter as indicator of resonance

(d) direct reading calibration with no confusing production of beat notes or harmonics.

The disadvantages are as follows.

(a) Only the 'order' of the frequency being checked is indicated. Thus an absorption frequency meter will show whether a transmitter frequency is nearer one end or the other of the amateur band, but not whether the frequency is just inside or just outside the limit of the band. The accuracy is about 5–10%.
(b) It lacks sensitivity. To obtain an indication in low-power transmitters, or in receiver tuned circuits, the absorption frequency meter must be held close to the coil being checked. Screening or other components may make this difficult, if not impossible.
(c) The presence of the frequency meter may cause an appreciable de-tuning of the circuit under test.

An absorption frequency meter is useful for checking that frequency-multiplying stages are tuned to the correct harmonics, for example, in checking whether a particular stage being driven by a 7MHz oscillator is tuned to 21MHz (third harmonic) or to 28MHz (fourth harmonic). Checking of harmonics is necessary in the initial tuning up of VHF equipment, which may operate on the 18th harmonic or higher of a given crystal frequency.

A rectifier and a microammeter can be added to the absorption frequency meter so that the meter itself can indicate resonance. This is shown in Fig 10.3. With the addition of a short antenna coupled to the coil, this arrangement becomes the so-called 'field strength meter' which may also be used for searching for parasitic oscillations in the transmitter. If a close-circuit jack for headphones is connected in series with the meter, the circuit can be used for monitoring an amplitude-modulated transmission.

The dip meter

This is more flexible than the absorption meter in that the tuned circuit under test does not have to be energised.

It consists of an oscillator using a FET. When the oscillator is tuned to the frequency of the tuned circuit under test, power is absorbed from the oscillator. This is indicated by a dip in the source current of the FET. The dip oscillator has the same

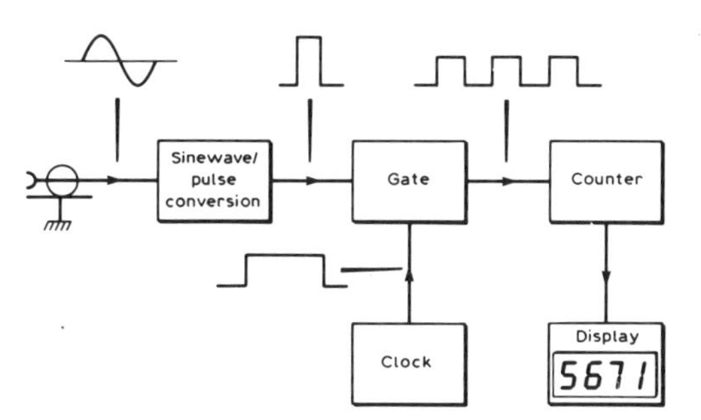

Fig 10.4. Simplified block diagram of digital frequency meter

accuracy as the absorption meter but has the disadvantage that a power source is needed.

The digital frequency meter

This type of frequency meter utilises integrated circuits to count electronically the number of complete sine waves in a given period. Although these digital integrated circuits are themselves complex, the principle of operation of the digital frequency meter (DFM) is quite simple; it consists of four major circuits, the 'clock', the 'gate', the 'counter' and the 'display'. Fig 10.4 is a simple block diagram of a DFM.

The clock produces a series of pulses of length 1s, 100ms, 10ms, 1ms and so on. These pulses are derived from a crystal-controlled 1MHz (or 5MHz) oscillator and their duration is controlled to a very high degree of accuracy. The pulses are applied to the gate which may be considered as an electronic on/off switch operated by the clock.

When the clock pulse opens the gate, the sinusoidal signal whose frequency is to be measured, having been converted to a train of square pulses, is allowed to pass to the counting circuits. As the name suggests, these count the number of pulses (cycles) which arrive while the gate is open and then pass the answer to the display circuits which present the answer in the form of a digital display.

The accuracy of the DFM depends upon the accuracy of the clock. If this produces pulses which are longer than they should be, the gate remains open for a longer period and more pulses are counted. If the clock pulse is too short, fewer pulses are counted. Accuracy can be increased if the clock oscillator crystal is housed in a crystal oven (thermostatically controlled) to hold the crystal temperature more or less constant.

The counting circuits will only respond to complete pulses and this may give rise to errors in measurement. For example, consider a signal of frequency of, say, 5671Hz. If the gate is open for only 0.1s, 567 whole pulses will be counted and it appears that the frequency is 5670Hz. If the gate is opened by the clock pulse for 0.01s, only 56 complete pulses are counted, suggesting that the frequency is 5600Hz and so on. Thus the duration of the clock pulse plays an important point in determining the accuracy of measurement.

In practice, the clock pulse duration or 'resolution' of a DFM can be set by a switch on the front panel of the instrument. With careful use an accuracy of ±5Hz can be achieved. This is more than adequate for most amateur applications.

Many types of IC may be used in digital frequency meters. The more common (and cheaper) types will permit measurement of frequencies up to 100MHz or so, and this may be extended by the latest ICs to well over 1000MHz.

The digital frequency meter is the most accurate type of frequency meter available.

The cathode-ray oscilloscope

The cathode-ray oscilloscope is probably the most valuable tool used in electronics. The heart of the oscilloscope is the cathode-ray tube. In this device, electrons emitted from an indirectly heated cathode are focused into a beam of small diameter which, when it strikes the front of the tube or screen, causes a special coating on the internal surface of the screen to fluoresce, creating a spot of light which may be blue or green according to the coating material used. On its way to the screen, the electron beam passes between one pair of parallel plates and then another pair at right-angles to the first. These are called the 'X and Y deflector plates'. If a voltage is applied between the deflector plates, the beam is deflected one way or the other according to the polarity of the voltage applied, and so the spot on the screen moves. Thus, if a voltage proportional to time is applied to one pair of plates, the horizontal-deflector (X) plates, and an alternating voltage is applied to the vertical-deflector (Y) plates, the spot traces out the waveform of the alternating voltage.

Timebase

The voltage applied to the X plates (ie horizontal motion) is generated inside the oscilloscope itself by what is usually referred to as the 'timebase'. The speed of the timebase can be varied over a large range to accommodate signals of widely differing frequencies.

For the less-expensive oscilloscope as used by the amateur, the range of the timebase is from 1µs/cm to 1s/cm.

Y amplifiers

The signals to be observed are usually very small. If they are applied directly to the Y plates, little, if any, movement in the vertical direction would be observed. It is necessary, therefore, to amplify the signals before applying them to the Y plates. This amplifier is called the 'Y amplifier' and is built into the oscilloscope. Several ranges of amplification are available, and each is calibrated so that a known voltage applied to the input causes a pre-determined deflection in the vertical direction. For amateur equipment the range of the Y amplifiers is from 5mV/cm to 100V/cm.

It should also be noted that the Y amplifiers have a limited bandwidth. Outside this bandwidth the calibration becomes less accurate.

Trigger

To get a stationary display on the oscilloscope it is necessary to start the timebase at the same point on the waveform for

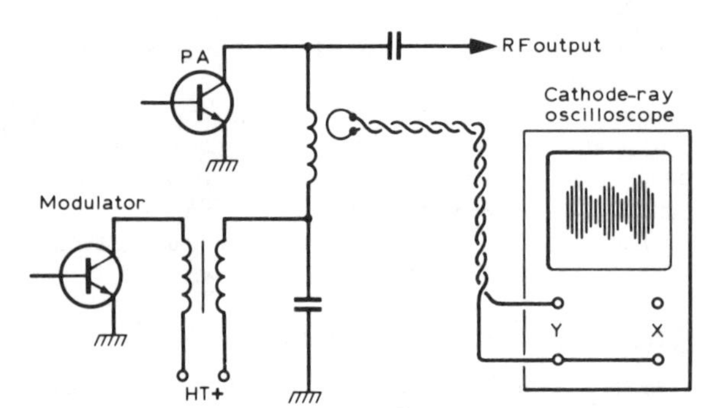

Fig 10.5. Arrangement for measuring modulation depth in which waveform of modulated carrier is displayed. Linear oscilloscope timebase is necessary

every sweep. This is accomplished by the 'trigger' circuits, the 'trigger level' control performing this function. On some older oscilloscopes stabilisation is accomplished by altering the frequency of the timebase until it synchronises with the input waveform; such oscilloscopes have a 'sync' control.

Measurement of modulation depth

It is imperative to ensure that, when amplitude modulation is used, the modulation depth on peaks does not exceed 100% as over-modulation creates serious interference (see Chapter 8).

The actual depth of modulation may be measured by displaying the waveform of the modulated output of the transmitter on an oscilloscope.

The circuit arrangement is shown in Fig 10.5 which is largely self-explanatory. Typical patterns produced are shown in Fig 10.6. By measuring the height R corresponding to a modulation peak, and the height of the unmodulated carrier (S) the depth of modulation can be calculated directly:

$$M = \frac{R - S}{S} \times 100\%$$

SWR measurement

As a result of the almost inevitable mismatch of an RF transmission line or feeder, neither the voltage nor the current is constant along the length of the line, that is, the voltage and current will vary according to the point along the line at which they are measured.

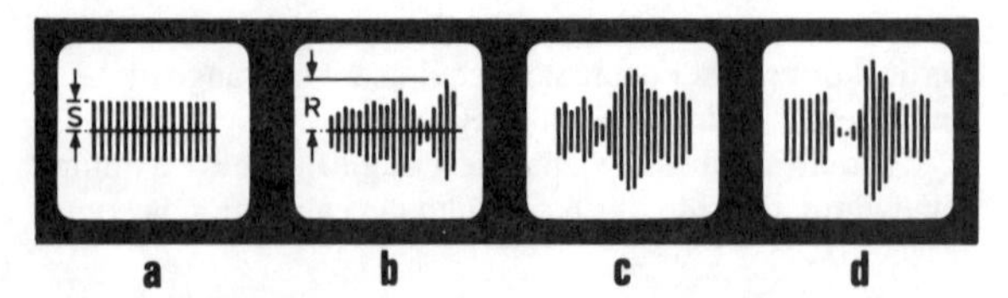

Fig 10.6. Typical patterns obtained by method shown in Fig 10.5: (a) unmodulated carrier; (b) modulation depth, 50%; (c) depth, 100%; (d) over-modulation. Break-up of carrier shows that over-modulation is occurring

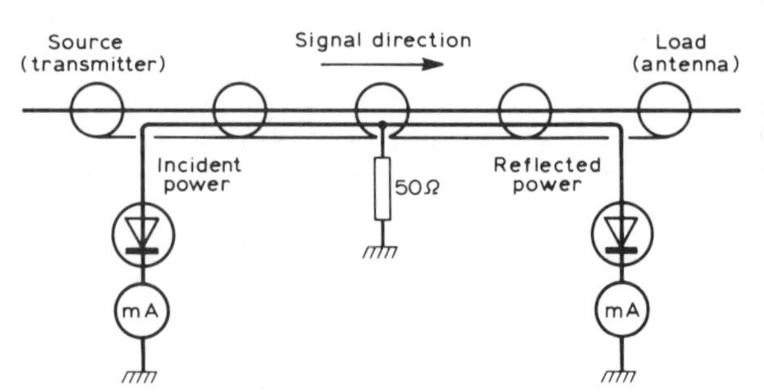

Fig 10.7. Basic reflectometer circuit

The performance of an RF transmission line can therefore be judged only by the measurement of its standing wave ratio (SWR) as defined in Chapter 7.

SWR is measured by means of a circuit known as a 'reflectometer'. The principle of operation is that when an RF transmission line is terminated in an impedance other than the characteristic impedance, ie when the line is not properly matched, part of the signal is reflected back long the line. The ratio of the power reflected to the power incident on the termination is directly related to VSWR.

The basic reflectometer circuit is shown in Fig 10.7. An insulated wire is threaded down between the outer shield and the insulation of a coaxial cable. When an RF signal passes down the coaxial cable part of it is coupled into the wire. This RF signal is rectified and the resulting DC is displayed on a milliammeter. This is an indication of the 'forward' power. A reflected signal that is travelling from termination to source will also be coupled into the wire. However, it will do so in such a way that it is rectified by the other diode and displayed by the other milliammeter. This is the 'reflected' power. SWR meters can be made easily from published data. These generally use a somewhat different circuit which avoids the centre tap on the wire threaded down the coaxial cable. A single meter is usually switched to read forward or reflected power.

Measurement of power of an SSB transmitter

The output power and hence the current drawn by the output stage of an SSB transmitter varies at a syllabic rate, ie in accordance with the speech waveform. Measurement of input power is therefore not possible because the conventional meter is much too slow in operation to follow the rapid variation of the input current.

The power rating of an SSB transmitter is therefore expressed as peak envelope power (PEP) as derived in Chapter 4. This is the power which exists at the peaks of the speech waveform. The maximum permitted PEP is 400W (26dBW).

The recommended method of measuring PEP is to monitor the output of the transmitter on a dummy load, by means of a cathode-ray oscilloscope, when the transmitter is modulated by the output of a 'two-tone generator' as shown in Fig 10.8. This device contains two AF oscillators which produce non-harmonically related sinusoidal tones of equal amplitude which are combined.

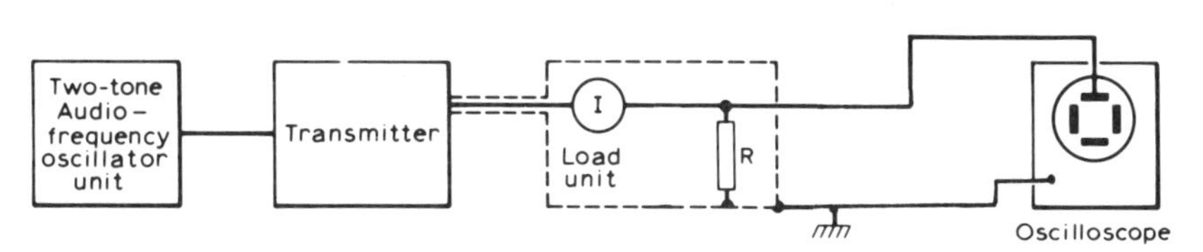

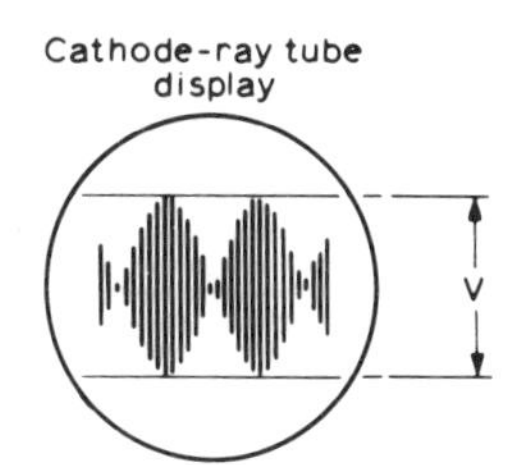

Fig 10.8. Method of measuring PEP output in relation to mean output with two-tone source. Power output = *I*I*R*; peak envelope power = 2 × mean power represented by *V*

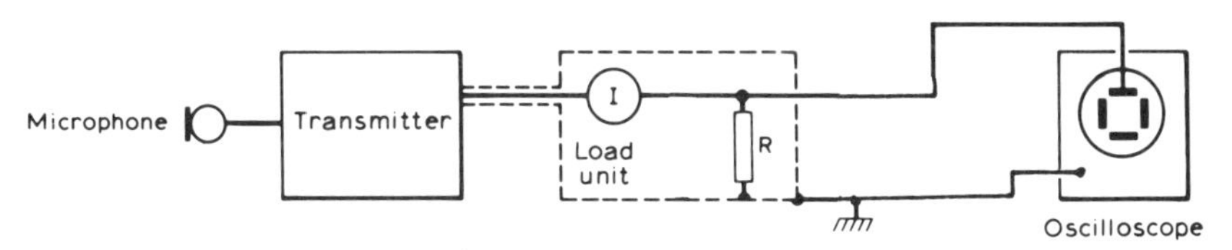

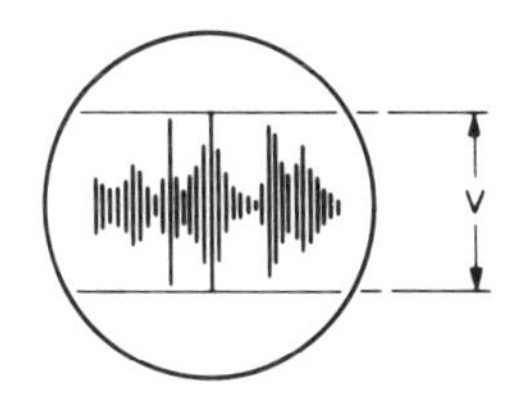

Fig 10.9. Speech peaks should not exceed level determined by *V* in two-tone test

The transmitter operating level is set to produce a mean power output in the dummy load of 200W as measured by the current flowing through it. The pattern produced on the oscilloscope will be as in Fig 10.8 and the limits of the deflection V should now be marked by two thin lines drawn by a Chinagraph pencil. The mean power as just measured (200W) is equivalent to a PEP of twice this value, ie 400W. Thus the two lines on the 'scope face represent the deflection which corresponds to an output of 400W PEP. It is not essential to set the transmitter output to 200W mean, any power will do, but 200W is a convenient level because it causes a deflection which corresponds to the maximum PEP permitted.

To avoid waveform errors, the current in the load resistor is measured by a thermocouple meter. If an RF voltmeter is used to measure the voltage across the load, waveform errors in the voltmeter must be taken into account in calculating the power output.

When high-power equipment is used, the overall accuracy of the power measurement must be known and taken into account in calculating the power in order to avoid exceeding the licensed power.

When the two-tone generator is replaced by the microphone, the oscilloscope now shows the extremely peaky speech waveform (see Fig 10.9). The maximum deflection must not be allowed to exceed that deflection which corresponds to 400W PEP.

Other power levels may be determined as follows. Suppose the deflection corresponding to 400W PEP is 5cm and the deflection resulting from peaks of speech is 2.5cm. The PEP is then:

$$400\text{W} \times \frac{2.5^2}{5^2} = 400\text{W} \times \left(\frac{2.5}{5}\right)^2$$

$$= 400\text{W} \times (1/2)^2$$

$$= 100\text{W}$$

Note that the deflections are squared. This is because the deflection on the oscilloscope is proportional to the voltage which causes it, whereas power is proportional to the square of the voltage.

It will be appreciated that the majority of SSB transmitters and transceivers have outputs of 200W or less. The necessity for PEP measurement only really arises when a high-power transmitter is in use, or where a transceiver is followed by a linear amplifier, many of which are rated at power levels greater than that permitted by the UK licence.

The procedure described above and illustrated in Fig 10.8 can be used to measure the power output of a transmitter operating in any mode. A single tone of a convenient frequency (say 1000Hz) should be used to modulate an AM transmitter. The modulation depth should be adjusted to 100% by variation of the tone input to the modulator when the transmitter is operating at the input intended. If CW or FM is in use, the transmitter should be in the 'key down' or 'transmit' state respectively.

Power meters

The 'Monimatch' SWR meter shown in Fig 10.7 is a simple and reasonably satisfactory instrument. The main disadvantage is that the maximum meter deflection is directly proportional to the frequency at which it is being used.

A more effective circuit uses a broad-band current transformer as a sensor. The secondary winding of this is a few turns of wire wound on a small toroidal ferrite core. The single-turn primary is a short length of coaxial cable through the toroidal core. This device can be easily made at home.

The advantage of this circuit is that the deflection produced is independent of the frequency at which it is used. Such a meter can also be calibrated to read power (forward and reverse) and, by the inclusion of a long-time-constant circuit, peak envelope power (PEP) as well.

Errors in measurement

When taking any measurements, the effect of inaccuracies in the meter used, ie the tolerances (which may be positive or negative) on the meter readings must be taken into account.

To take a simple case, suppose the input to a PA is measured as 15V at 10A, ie 150W. If both meters are reading low by say 5%, the actual input is (15V + 5%) = 15.75V at (10A + 5%) = 10.5A or 165W. This can be considered as a worst case as it is unlikely that both meters will read low by this amount. The error in reasonably new good-quality meters should be less than 5%, while old meters of unknown history may be in error by more than 10%.

Tolerances are usually expressed as a percentage (eg ±2%), a value (eg ±0.3V or ±200Hz) or as so many 'parts per million' (eg ±100ppm). A number of tolerances affecting a reading may be added together, although this is likely to give a pessimistic total. As an example, consider the following sample examination question:

A transmitter operating in the band 21MHz to 21.450MHz has a frequency tolerance of 100 parts in one million and a radiated bandwidth of 6kHz when using emissions of type A3E. If the frequency checking equipment at the station has a frequency tolerance of 10 parts in one million, what is the lowest frequency a licensee can use that ensures no emission below 21MHz?

(a) 21,005.1kHz
(b) 21,053.1kHz
(c) 21,008.31kHz
(d) 21,005.31kHz

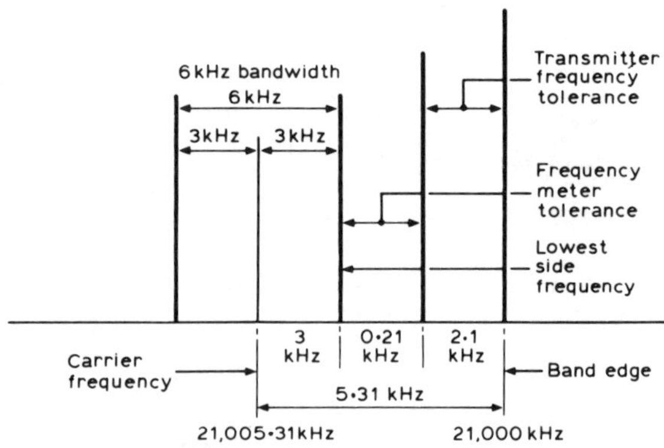

Fig 10.10. Diagrammatic solution of frequency-tolerance problem

The tolerance build-up is as follows:

Band edge	21,000.0kHz
Frequency tolerance is 100ppm, ie 100Hz per megahertz	2.1kHz
Frequency meter tolerance 10ppm	0.21kHz
A3E bandwidth is 6kHz, ie carrier must be in centre of this band	3.0kHz
	21,005.31kHz

and thus the lowest frequency which can be used is 21,005.31kHz. This is shown diagrammatically in Fig 10.10.

The DTI recommendations on frequency measurement in the amateur station in the booklet *How to become a Radio Amateur* should be studied.

CHAPTER 11

Operating practices and procedures, repeaters and satellites

Telegraphy

Effective communication by telegraphy implies the use of internationally agreed symbols and abbreviations so that difficulties arising from language differences are eliminated to a fairly large extent.

The symbols and abbreviations used in amateur radio are based on the international Q-code and the procedures used in marine and commercial radio telegraphy. Many other abbreviations are based on phonetic English.

The international Q-code is a series of questions and answers, and the Q-signals in Table 11.1, which are taken from the official list, are commonly used in the amateur service. Amateurs use many of the Q-signals as nouns as well as in question-and-answer form (see Table 11.2).

The most common procedure and punctuation signals are shown in Table 11.3.

Reports on readability and signal strength (and tone in the case of telegraphic signals) are given in terms of the RST code (see Table 11.4).

Table 11.1. International Q-code (extract)

QRG	Will you tell me my exact frequency? Your exact frequency is ....kHz
QRH	Does my frequency vary? Your frequency varies.
QRI	What is the tone of my transmission? The tone of your transmission is ....(amateur T1–T9).
QRK	What is the readability of my signals? The readability of your signals is......(amateur R1–R5).
QRL	Are you busy? I am busy. Please do not interfere.
QRM	Are you being interfered with? I am being interfered with.
QRN	Are you troubled by static? I am troubled by static.
QRO	Shall I increase power? Increase power.
QRP	Shall I decrease power? Decrease power.
QRQ	Shall I send faster? Send faster.
QRS	Shall I send more slowly? Send more slowly.
QRT	Shall I stop sending? Stop sending.
QRU	Have you anything for me? I have nothing for you.
QRV	Are you ready? I am ready.
QRX	When will you call me again? I will call you again at ....hours.
QRZ	Who is calling me: You are being called by .... (on kHz).
QSA	What is the strength of my signals? The strength of your signals is..........(amateur S1–S9).
QSB	Are my signals fading? Your signals are fading.
QSD	Is my keying defective? Your keying is defective.
QSL	Can you give me acknowledgement of receipt? I give you acknowledgement of receipt.
QSO	Can you communicate with ....direct or by relay? I can communicate with ....direct (or by relay through..........).
QSP	Will you relay to........? I will relay to ....
QSV	Shall I send a series of VVVs? Send a series of VVVs.
QSY	Shall I change to another frequency? Change to transmission on another frequency (or on .....kHz).
QSZ	Shall I send each word more than once? Send each word twice?
QTH	What is your location? My location is..........
QTR	What is the correct time? The time is ....hours.

Speed of sending

The golden rule is 'never send at a greater speed than you are able to receive'. The speed of sending should depend to a large extent on circumstances: when conditions are poor with low signal strengths or in heavy interference it is sensible to send more slowly. It is a good principle to send at the same speed as the operator at the other end.

The good CW operator is the one whose copy is easy to read and who does not send faster than he is capable of doing properly. An indifferently sent 25 words per minute may well be almost unreadable but a properly sent 25wpm is considerably easier to read than a badly sent 20wpm.

It is probably true to say that most telegraphy on the amateur bands is at a speed of somewhat less than 20wpm.

Establishing communication

The first step is to spend a short time listening on the band it is proposed to use in order to check:

(a) if conditions are good or bad;

(b) who is working who and what signal reports are being exchanged.

Table 11.2. Use of Q-signals as nouns

QRA	Address	QRP	Low power
QRG	Frequency	QRT	Close down
QRI	Bad note	QRX	Stand by
QRK	Signal strength	QSB	Fading
QRM	Interference from other stations	QSD	Bad sending
		QSL	Verification card
QRN	Interference from atmospherics or local electrical apparatus	QSO	Radio contact
		QSP	Relay message
		QSY	Change of frequency
QRO	High power	QTH	Location

Table 11.3. Punctuation/procedure signals

Punctuation	
Question mark	di-di-dah-dah-di-dit
Full stop	di-dah-di-dah-di-dah
Comma*	dah-dah-di-di-dah-dah

* Sometimes used to indicate exclamation mark

Procedure signals	
Stroke (/)	dah-di-di-dah-dit
Break sign (=)	dah-di-di-di-dah
End of message (AR)**	di-dah-di-dah-dit
End of work (VA)**	di-di-di-dah-di-dah
Wait (AS)**	di-dah-di-di-dit
Error	di-di-di-di-di-di-di-dit
Invitation to transmit (general) (K)	dah-di-dah
Invitation to transmit (specific station) (KN)**	dah-di-dah-dah-dit

** AR, VA, AS and KN are sent as one character

Table 11.4. The RST code

Readability	
R1	Unreadable
R2	Barely readable, occasional words distinguishable
R3	Readable with considerable difficulty
R4	Readable with practically no difficulty
R5	Perfectly readable
Signal strength	
S1	Faint, signals barely perceptible
S2	Very weak signals
S3	Weak signals
S4	Fair signals
S5	Fairly good signals
S6	Good signals
S7	Moderately strong signals
S8	Strong signals
S9	Extremely strong signals
Tone	
T1	Extremely rough hissing note
T2	Very rough AC note, no trace of musicality
T3	Rough, low-pitched AC note, slightly musical
T4	Rather rough AC note, moderately musical
T5	Musically modulated note
T6	Modulated note, slight trace of whistle
T7	Near DC note, smooth ripple
T8	Good DC note, just a trace of ripple
T9	Purest DC note

If the note appears to be crystal-controlled add X after the appropriate number. Where there is a chirp add C, drift add D, clicks add K.

The propagation forecasts published each month in *Radio Communication* are invaluable as an indication of the part of the world likely to be heard at a particular time.

There are two ways of establishing communication:

(a) by calling a specific station;
(b) by transmitting a 'CQ' ('general invitation to reply') call.

On most bands there are always many stations to be heard calling 'CQ' so it is generally preferable to answer such a call rather than to initiate another. However, a CQ call made when a band appears 'dead', particularly on 21 or 28MHz, sometimes results in an unexpected contact.

Calling procedures

Calling a specific station

First of all, 'net' on to the frequency of the station it is proposed to call, ie adjust transmitter frequency onto the signal being received. The basic call is:

G7AA G7AA G7AA de G7ZZ G7ZZ G7ZZ KN

'de' here means 'from'. Note that 'KN' is an invitation to a specific station, G7AA, and no other, to reply.

This basic call may be varied, ie in poor conditions or heavy interference it would be advisable to send the station call (G7ZZ) five or six times.

Initiating a CQ call

The first step is to choose a frequency where no other station is operating – this is often very difficult and it has to be accepted that virtually every signal is likely to cause some interference to another station somewhere. Before initiating a CQ, the signal 'QRL?' is often sent two or three times; this is intended to enquire "is this frequency in use?" If so, then the answer given is 'QRL' meaning, "I am using this frequency" or sometimes 'pse QSY' meaning "please change frequency". The action is then obvious! The basic call is:

CQ CQ CQ de G7ZZ G7ZZ G7ZZ K

Note here that 'K' is a general invitation to any station to reply.

Basic calls may be repeated up to five or six times – this depends on conditions and activity on the band. Never send a long series of CQ signals without interspersing the station call.

A CQ call may be made specific, eg

CQ DX CQ DX CQ DX de G7ZZ G7ZZ G7ZZ K

or directional, eg

CQ VK CQ VK CQ VK de G7ZZ G7ZZ G7ZZ K

Prolonged CQ calls or calls to a specific station should be avoided. These cause unnecessary interference, particularly to stations in the immediate vicinity who generally accept interference from a local station in contact but who will not take too kindly to continuous CQ calls. An unanswered CQ is therefore best followed by a short period of listening.

If a station replying cannot be positively identified due to low signal strength or interference, the signal 'QRZ?' ("who is calling me?") may be used, eg

QRZ? QRZ? de G7ZZ G7ZZ G7ZZ KN

Note that KN and not K is now used. A 'QRZ?' call should be brief and not extended as an additional CQ call. 'QRZ?' is not an alternative to 'CQ'.

Having established communication a CW contact (QSO) is likely to follow the general form shown here.

G9AA G9AA DE G7AA G7AA = GA OM ES MNI TNX FER CALL = AM VY PSED TO QSO U = UR SIGS RST 579 = QTH IS LONDON = NAME IS GEO = HW? AR G9AA DE G7AA KN

These transmissions are known as 'overs'. The following should be noted.

(a) Report, QTH and name are generally sent twice (but not more than three times).
(b) Each sentence is separated by the break sign (dah-di-di-di-dah).
(c) Each callsign sent twice at the beginning is generally adequate but may be sent once only or three times depending on signal strength and interference etc.
(d) Each callsign is sent once only at the end – there is no point in sending them more than once.
(e) Names may just as well be abbreviated.
(f) 'HW?' means "how do you receive me?"
(g) 'AR' signifies "end of message".
(h) 'KN' means specific station (G9AA) to reply.

The contact may then continue:

G7AA G7AA DE G9AA G9AA = R ES VY GA GEO = GLD TO QSO AGN = THINK WE QSO LAST YEAR ON 3R5? = MNI TNX FER RPRT = UR RST 569 QSB QRM = QTH IS BIRMINGHAM = NAME IS MAC = RIG IS FT101 AT 150 WATTS INPUT = ANT IS TRAP DIPOLE = WX IS COLD ES DULL = OK? AR G7AA DE G9AA KN

The following should be noted:

(a) 'R' signifies "received all sent", which is obviously preferable to such phrases as "solid cpy hr" etc.
(b) '3R5' – here 'R' indicates a decimal point, ie 3.5.

The contact then goes on:

G9AA G9AA DE G7AA G7AA = MOST OK MAC BD QRM AT END = TNX FER RPRT ES INFO ON RIG = HR HOME MADE TX ES INPUT IS 120 WATTS = RX IS AR88D = ANT IS 132 FT END FED = YES WE DID QSO LAST YEAR = CONDX FB FER DX BUT EU STNS VY STRONG = QRU? AR G9AA DE G7AA KN

G7AA G7AA DE G9AA G9AA = R ES TNX ALL = MNI TNX QSO ES HPE CUAGN SN = PSE QSL VIA BURO = VY 73 ES DX = GB OM ES GL AR G7AA DE G9AA KN

Note that this is G9AA's last transmission but the contact has not finished because he is about to receive G7AA's final over. G9AA therefore finishes with 'KN' and not 'VA'.

G9AA G9AA DE G7AA G7AA = R FB SIGS NW OM = QRT = MNI TNX QSO ES HPE CUAGN = QSL OK VIA RSGB = 73 ES DX = CHEERIO MAC ES ALL THE BEST AR G9AA DE G7AA VA CL.

This is the last transmission of the contact and G7AA therefore uses 'VA', meaning "I have finished". 'CL' indicates that G7AA is closing down.

The above describes a fairly basic contact but in practice CW contacts range from just an exchange of RST/QTH/name to a chat lasting an hour or more.

Table 11.5. Amateur abbreviations (CW)

AA	All after...(used after a question mark to request a reception)
AB	All before...(see AA)
BK	Signal used to interrupt a transmission in progress
CFM	Confirm (or I confirm)
NIL	I have nothing to send you
NW	Now
OK	We agree (or it is correct)
QSLL	If you send a QSL, I will do likewise
RPT	Repeat (or I repeat)
TFC	Traffic
W	Word(s)
WA	Word after (see AA)
WB	Word before

'K', 'KN' and 'VA' are probably the most misused symbols in amateur radio. Remember: 'K' is an invitation to any station to reply. 'KN' is an invitation to a specific station station to reply. 'VA' means "I have finished".

When tuning across a band, if only 'G9AA K' is heard, one is entitled to call G9AA. If 'G9AA KN' is heard, it means that G9AA is in contact with someone else or has just called someone, and one should therefore not call G9AA. On the other hand if 'G9AA VA' is heard it should indicate that G9AA has just finished a contact and therefore one is entitled to call him.

It follows that if these symbols are not used correctly a considerable amount of annoyance can be created.

At the conclusion of a contact always listen for a few seconds – someone may be calling you! Do not immediately send 'QRZ?' after sending 'VA' unless you suspect someone may be waiting for you. A very short CQ call, ie just 'CQ DE G9AA K' will indicate that you are now ready to accept a call.

The procedure of 'tail-ending' means a very short call, ie 'DE G9AA KN', to a station immediately he has finished his last transmission. It is most effective when both sides of the contact have been heard but it does require that 'VA' is correctly used. It is accepted by some good operators but is open to abuse and likely to cause irritation if used carelessly. It should be avoided until a fair amount of operating skill has been acquired.

Do not 'send double' unless specifically asked to do so or your signals have been reported as, say RST 339.

Abbreviations used in amateur telegraphy are understood throughout the world. Many of these are given at the end of this chapter.

Telephony

Whereas a poor or inconsiderate CW operator is a nuisance only to his fellow amateurs, bad telephony operation discredits amateur radio generally. Our hobby can be too easily judged by the quality of our telephony transmissions, the subjects discussed and the procedures used.

The major portion of amateur radio traffic is now carried out using telephony. Though this mode does not require the knowledge of codes and abbreviations, correct operation is

Table 11.6. Informal amateur abbreviations (CW)

ABT	about	ES	and	INPT	input	SSB	single sideband
ADR	Address	FB	fine business	LID	poor operator	STN	station
AF	africa	FM	frequency modulation	MNI	many	SUM	some
AGN	again			MOD	modulation	SWL	short-wave listener
ANI	any	FER	for	MSG	message	TKS	thanks
ANT	antenna	FONE	telephony	MTR	meter (or metres)	TMW	tomorrow
BCNU	be seeing you	FREQ	frequency	NA	North America	TNX	thanks
BD	bad	GA	go ahead, or good afternoon	NBFM	narrow band frequency modulation	TRX	transceiver
BFO	beat frequency oscillator					TVI	television interfer- ence
		GB	goodbye				
BK	break-in	GD	good day	NR	number	TX	transmitter
BLV	believe	GE	good evening	OB	old boy	U	you
BUG	semi-automatic key	GLD	glad	OM	old man	UR	your
CK	check	GM	good morning	OP	operator	VY	very
CLD	called	GN	good night	OT	old timer	W	watts
CONDX	conditions	GND	ground (earth)	PSE	please	WID	with
CRD	card	GUD	good	PWR	power	WKD	worked
CUD	could	HAM	amateur transmitter	RPRT	report	WKG	working
CUAGN	see you again	HI	laughter	RX	receiver	WL	will or well
CUL	see you later	HPE	hope	SA	South America	WUD	would
CW	continuous wave	HR	hear or here	SED	said	WX	weather
DR	dear	HRD	heard	SIG	signal	XMTR	transmitter
DX	long distance	HV	have	SKED	schedule	XYL	wife
ELBUG	electronic key	HVY	heavy	SN	soon	YL	young lady
ENUF	enough	HW	how	SRI	sorry	73	best regards

more difficult than it may appear at first sight, as is only too apparent after listening on any amateur band.

Part of the problem is that many operators will have acquired some bad habits in their pronunciation, intonation and phraseology even before entering amateur radio. To these are then added a whole new set of cliches and mannerisms derived from listening to bad operators. Some of these can be extremely difficult to remove once learnt, even if a conscious effort is made.

Conversation

It is important to speak clearly and not too quickly, not just when talking to someone who does not fully understand the language, but at all times.

The use of CW abbreviations (including 'HI') and the Q-code should normally be avoided. The Q-code should only be used on telephony when there is a language difficulty.

Plain language should be used, and cliches and jargon should be kept to a minimum. In particular, avoid the use of "we" when "I" is meant and "handle" when "name" is meant. Other silly habits include saying "that's a roger" instead of "that's correct", and "affirmative" instead of "yes". The reader will no doubt have heard many more. Taken individually each is almost harmless, but when combined together give a false-sounding 'radioese' which is actually less effective than plain language in most cases.

The phonetic alphabet listed in the amateur licence is in Table 11.7, and should be used only when necessary to clarify a callsign, or the spelling of a word or in bad reception conditions. Never change the alphabet in mid-stream, ie do not say "Alpha Papa Juliett", followed by "America, Pacific, Japan" because this can be confusing. Similarly "Alpha Alpha Papa Papa Juliett Juliett" does not really help. Do remember, if callsigns, name and location were spoken slowly and clearly, the majority of the 'rapid-fire' phonetic alphabets heard on the bands would be unnecessary.

Unlike CW operation, it is very easy to forget that the conversation is not taking place down a telephone line. Unless duplex operation is actually in use, the listening station cannot interject a query if something is not understood, and cannot give an answer until the transmitting station has finished. The result is often a long monologue, in which the listening station has to take notes of all the points raised and questions asked if a useful reply is to be given. This should not be necessary if these points are dealt with one at a time.

Procedure

As noted earlier, when calling a specific station it is good practice to keep calls short and to use the callsign of the station called once or twice only, followed by one's own callsign

Table 11.7. Recommended phonetic alphabet

A	Alpha	J	Juliett	S	Sierra
B	Bravo	K	Kilo	T	Tango
C	Charlie	L	Lima	U	Uniform
D	Delta	M	Mike	V	Victor
E	Echo	N	November	W	Whiskey
F	Foxtrot	O	Oscar	X	X-ray
G	Golf	P	Papa	Y	Yankee
H	Hotel	Q	Quebec	Z	Zulu
I	India	R	Romeo		

pronounced carefully and clearly at least twice using the phonetic alphabet, for example:

"WD9ZZZ. This is Golf Four Zulu Zulu Zulu calling, and Golf Four Zulu Zulu Zulu standing by."

Emphasis should be placed on the caller's own callsign, and not on that of the station called. If there is no response, the caller's callsign may be repeated once more after a brief listen.

As in CW operation, CQ calls should also be kept short and repeated as often as desired. An example would be:

"CQ, CQ, CQ, CQ. This is Golf Four Zulu Zulu Zulu calling, Golf Four Zulu Zulu Zulu calling CQ and standing by.'

There is no need to say which band is being used, and certainly no need to add "for any possible calls, dah-di-dah!" or "K someone please" etc!

When replying to a call both callsigns should be given clearly, so that the calling station can check its callsign has been received correctly. From then on it is not necessary to use the phonetic alphabet for callsigns until the final transmissions. An example would be:

"Whiskey Delta Nine Zulu Zulu Zulu. This is Golf Four Zulu Zulu Zulu."

Once contact is established it is only necessary to give one's callsign at the intervals required by the licensing authority. A normal two-way conversation can thus be enjoyed, without the need for continual identification. If necessary the words "break" or "over" may be added at the end of a transmission to signal a reply from the other station. In good conditions this will not normally be found necessary. When FM is in use it is self-evident when the other station has stopped transmitting and is listening, because the carrier drops.

The situation is more complex where three or more stations are involved, and it is a good idea to give one's own callsign briefly before each transmission, for example:

"From G4ZZZ . . ."

At the end of the transmission the callsign should again be given, together with an indication of whose turn to speak it is next, for example:

". . . WD9ZZZ to transmit. G4ZZZ in the group."

It is not necessary to run through a list of who is in the group, and who has signed off (and who may possibly be listening) after each transmission, although it may be useful to do this occasionally.

Signal reports on telephony are usually given as a single two-digit number, in a similar fashion to the three-digit CW RST code.

It is recommended that both callsigns be given in the final transmission using the phonetic alphabet so that listening stations can check that they have them correct before calling, for example:

". . . This is Golf Four Zulu Zulu Zulu signing clear with Whiskey Delta Nine Zulu Zulu Zulu and Golf Four Zulu Zulu Zulu is now standing by for a call."

Note that some indication to listening stations is useful to indicate what is planned next. Such an indication is also appropriate if an immediate change to another frequency is intended, for example:

". . . This is Golf Four Zulu Zulu Zulu signing clear with Golf Two X-Ray Yankee Zulu Mobile. Golf Four Zulu Zulu Zulu now monitoring S20 for a call."

The restrictions imposed by the UK licence conditions are particularly relevant in telephony operation. It should be made a golden rule never to discuss politics, religion or any other matter which may offend the person to whom one is talking or anyone who may be listening.

Band planning

On most bands, certain sections are set aside for use by particular modes. The object of this 'band planning' is to ease communication by separating incompatible modes of transmission as far as possible so that everyone can get on with the business of communicating by his or her chosen method with the minimum of interference.

In some countries observance of band plans is mandatory, in others (eg USA) their use is determined by the class of licence held by the operator. They are not mandatory in the UK, but they should be observed at all times.

Comprehensive plans for all UK amateur bands from 1.8MHz to 47GHz are given in the RSGB *Amateur Radio Call Book*, which is published annually in October, and also in *Radio Communication* (February or March issue). It is necessary for the RAE candidate to know the reasons for, and the advantages of, band planning, and when licensed he must be familiar with the plans for the bands on which he is active. All band plans are subject to change, and the most recent versions should be consulted.

General

The following comments are equally applicable to both telegraphic and telephonic operation.

Honest reports, particularly on tone, should always be given. Do not give a report of RS(T) 59(9) to a station merely because he has just given you 59(9) or because you want his QSL card! S-meter readings should be treated with reserve, ie RST 519 is almost meaningless. (Awards or certificates of operating proficiency require a minimum signal report of RST 339/RS 33 – these represent just about the minimum usable signals).

The present, almost universal, use of reports of 599 (sent as 5NN) or 59 in contests and short 'rubber stamp' contacts should be frowned upon (unless of course the signals really are 59!).

Directional CQ calls should always be respected – G9AA would be considered a poor operator if he is heard calling an Australian station which has just transmitted a 'CQ USA' call.

AMATEUR RADIO STATION LOG

DATE	TIME (UTC) start	TIME (UTC) finish	FREQUENCY (MHz)	MODE	POWER (dBW)	STATION called/worked	REPORT sent	REPORT received	QSL sent	QSL rcvd	REMARKS
2 Nov '88	0800	0810	3	J3E	20	GM5ABC	59^{+10}	59^{+5}			Bert
"	0811	0820	145	F3E	16	G7XYZ	57	56			Terry first G7
"	0825	0830	14	J3E	20	CQ					No reply
"	1725	1735	145	F2D	16	GB7XYZ					Local packet mailbox
"	1740		Station closed down								
4 Nov '88	1030	/P	from 73 Antenna Lane, Squelch-on-Sea								
"	1031	1036	50	J3E	10	GIØXYZ	55	56			Jim, Bridgetown
"	1036	1045	50	J3E	10	G7XYL	58	58			Anne, Nr Squelch-on-Sea. QRM
"	1205	1215	433	F3E	13	G2XYZ	46	47			
"	1220		Station closed down								
5 Nov '88	0945	/P	from 73 Antenna Lane, Squelch-on-Sea								
"	0950	1005	144	J3E	16	GB2GUY	56	56	✓		Catherine Fawkesville
"	1010	1015	144	A1A	16	GD5ZZZ	542	541	✓		QSB! QSL via WF9XYZ
"	1526	1530	144	A1A	16	G7CW	579	589			Good keying!
"	1535		Station closed down and dismantled								
7 Nov '88	1810	1902	435	C3F	10	G7ZZZ	P3	P3			Ted, first ATV contact!
"	1930	1945	21	J2B	16	VK2ABC	559	569	✓		RTTY, Sid at Bandedge
"	1946	2005	21	J2B	16	ZL3222	569	559	✓		1st ZL on RTTY
"	2010		Test for TVI/Harmonic Radiation - Nothing noted.								
"	2020		Station closed down								
8 Nov '88	1735	1737	7	J3E	20	CQ					
	1738	1805	7	J3E	20	GIØZZZ	58	58			Nobby - chatted about G5RV ant
	1930	1945	51	F3E	10	GØSIX	55	55			Allen, wanted WAB ref.
	1950		Station closed down.								
NOTES											

Fig 11.1. Typical log entries

It is courteous to move off a frequency at the end of a contact if the station contacted was originally operating there.

The long-distance (DX) bands, particularly 14, 21 and 28MHz, should not be used for purely local contacts when these bands are open for long-distance working.

Do not call a station while its operator is in contact with someone else. Similarly, it is considered very poor operating to try to break in to a contact which is already taking place.

The various conditions of the amateur licence should be kept in mind as some of these have a bearing on operation.

Log keeping

Apart from the fact that a log of all transmissions is required, a well-kept log provides a record of contacts and friendships made, reports, conditions and other information on which applications for operating awards can be made.

The basic requirements of the Department of Trade and Industry with regard to log keeping are defined in the licence conditions as follows:

1. A permanent record must be kept in one book (not loose leaf).
2. The following data must be recorded:
   Date.
   Time of commencement of operation.
   Class(es) of emission.
   Transmitter power
   Frequency band(s).
   CQ calls.
   Callsigns of stations called and with whom communication is established.
   Time of establishing and ending communication with each station.
   Tests carried out (eg EMC).
   Time of closing the station.

The detailed requirements for log keeping will be found in *How to become a Radio Amateur*. Check these before taking the examination.

No particular method of recording the date is specified but it should be noted that '10/1/88' means '10 January 1988' in the UK and most of the world but '1 October 1988' in the USA.

Fig 11.1 illustrates a number of typical log entries which satisfy the basic licence requirements.

Repeaters and satellites

The past few years have seen the introduction of two new aspects of the amateur service: repeaters and satellites. Both of these help the VHF and UHF operator to increase his or her range, although by entirely different means.

Repeaters

As the name suggests, these are intended to receive (on the 'input channel') VHF or UHF signals from portable and mobile stations, and re-transmit (relay) them on a different frequency (the 'output channel') within the same amateur band. Input and output frequencies are spaced by 600kHz at 144MHz and 1.6MHz at 433MHz.

Repeaters are unmanned and entirely automatic in operation. They are situated at the top of a hill or where the antenna can be positioned as high as possible. Thus the rather limited range of portable and mobile equipment may be increased from 10–15km to something like 50km. This is indicated in Fig 11.2. This increase in range is at no cost to the amateur in respect of extra power or improved antennas.

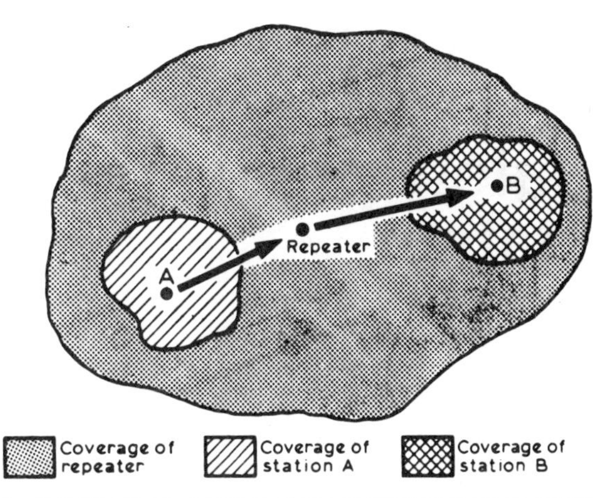

Fig 11.2. The improved range of communication between mobile stations when using a repeater. The simplex coverage areas of stations A and B shown are constantly changing shape as the two vehicles pass through different terrain, and thus station B would have to be quite close to station A before reliable communication was possible

Continuous operation of the repeater transmitter is undesirable and therefore means of remotely turning the repeater on and off is included; this is generally known as 'accessing' the repeater. In UK repeaters accessing is achieved by transmitting a 1750 ±25Hz tone (the 'toneburst'), which is approximately 0.5s long, at the beginning of each period of use of the repeater. (Some repeaters do not require a toneburst, being carrier operated, ie they are activated by the presence of a user's carrier on the input channel).

The tone switches on the transmitter and starts an internal clock which will turn the repeater transmitter off after typically one minute. This is known as 'time-out' and is intended to keep user transmissions short.

Repeaters operate only with frequency-modulated signals which must be of the correct frequency and deviation. The frequency and deviation are monitored continuously and if these parameters fall at any time below the standard required for valid access the repeater transmitter may be shut down.

When an 'over' is finished and the incoming signal disappears, the repeater will, after a short delay, indicate its readiness for another input transmission by transmitting either a 'K' or 'T' in morse code.

In the UK there are approximately 75 repeaters on the 144MHz band and 150 on 432MHz, and all use FM. A much smaller number of repeaters use other modes and frequencies. The current list can be obtained from the RSGB.

Satellites

A number of satellites specifically designed for amateur radio have been launched since 1961. These have been called Oscar 1, Oscar 2 etc (from Orbiting Satellite Carrying Amateur Radio).

The satellites are 'transponders' – they will accept CW, SSB or RTTY signals over a band of frequencies and re-transmit them in another amateur band. Current satellites have 432 to 144MHz, 144 to 432MHz, 144 to 28MHz and 1269 to 435MHz transponders.

Accessing a satellite does not require an access tone as the satellite monitors incoming signals continuously and its transmitter is permanently operational.

Obviously the satellite will only receive sufficient signal when it is in direct line-of-sight to the transmitter. This happens periodically when the satellite, which is rotating round the earth, appears over the horizon.

The power required to use the Oscar series transponders is 80–100W erp. This power level is usually achieved with a low-power transmitter (10–15W output) followed by an antenna with a gain of 10dB. The antenna must be pointed in the direction of the satellite to achieve this gain and so it must move in both azimuth and elevation in order to track the satellite properly. This can lead to complex antenna systems if best results are desired, although simple fixed antennas are capable of satisfactory operation if correctly designed. Information on the current amateur radio satellite scene can be obtained from the RSGB.

A brief comparison of the modes of operation of repeaters and satellites may be useful. A repeater accepts FM signals at a single frequency and re-transmits them in the same band. An access tone (1750Hz) is often required (sometimes only for the initial transmission of a contact) and the input signal must be of the correct frequency and deviation.

A satellite transponds a band of frequencies from one amateur band to another and will accept any form of modulation (CW, SSB and RTTY are preferred).

A repeater extends the user's range by a relatively small factor and the final range is still essentially local under most conditions. However, a satellite permits coverage of very long distances on VHF and UHF.

CHAPTER 12

Safety precautions in the amateur station

Probably the most important safety precaution in the amateur station arises from the use of high voltages. While much amateur equipment uses transistors operating from a 13.5V supply, many high-power HF, VHF and UHF linear amplifiers using valves operating at voltages over 2000V are in use.

It is not possible to define a voltage which is 'safe to touch' – no voltage source should ever be touched deliberately. The effects of an electric shock of a particular voltage depend on so many factors, from the source of the voltage to the state of health of the unfortunate person involved.

It is therefore best to consider any voltage above 50V as dangerous and to act accordingly. Probably the three most important precautions are:

1. Ensure that all equipment is satisfactorily earthed – the integrity of the earthing system should be checked periodically.
2. Switch off and disconnect from the mains supply before attempting the investigation or repair of any equipment.
3. Capacitors of high value, particularly if the dielectric is paper, will hold their charge for long periods (days/weeks). Bleed resistor chains should therefore be connected across high-voltage smoothing capacitors in order to dissipate this charge in a few seconds. Bleed resistor chains should be conservatively rated as regards dissipation and ideally should consist of two separate chains in parallel. In spite of the presence of bleed resistor chains, high-voltage smoothing capacitors should be shorted as recommended in paragraph 5 of the RSGB Safety Recommendations before any servicing of a power supply is carried out.

Using a low-voltage but high-current power supply must not be allowed to lull one into a state of false security. The input voltage is 240V, and this can be fatal under some circumstances. The high current capability may be sufficient to generate molten metal if the supply is accidentally shorted to earth by, say, a ring on one's finger.

The electromagnetic compatibility aspects of the protective multiple earthing system of house wiring are dealt with in Chapter 9. The installation of PME systems started in the mid 'seventies. It would seem likely therefore that the majority of houses in UK do not use PME. Before basing any grounding or safety precautions in your station on PME, it is essential to check with your local electricity board whether or not your house wiring uses PME.

The 'Safety Recommendations for the Amateur Radio Station' and 'Mobile Safety Recommendations' which follow should be studied carefully.

RSGB Safety Recommendations for the Amateur Radio Station

1. All equipment should be controlled by one master switch, the position of which should be well known to others in the house or club.
2. All equipment should be properly connected to a good and permanent earth. (Note A).
3. Wiring should be adequately insulated, especially where voltages greater than 500V are used. Terminals should be suitably protected.
4. Transformers operating at more than 100V rms should be fitted with an earthed screen between the primary and secondary windings.
5. Capacitors of more than 0.01µF capacitance operating in power packs, modulators, etc (other than for RF bypass or coupling) should have a bleeder resistor connected directly across their terminals. The value of the bleeder resistor should be low enough to ensure rapid discharge. A value of $1/C$ megohms (where C is in microfarads) is recommended. The use of earthed probe leads for discharging capacitors in case the bleeder resistor is defective is also recommended. (Note B). Low-leakage capacitors, such as paper and oil-filled types, should be stored with their terminals short-circuited to prevent static charging.
6. Indicator lamps should be installed showing that the equipment is live. These should be clearly visible at the operating and test position. Faulty indicator lamps should be replaced immediately. Gas-filled (neon) lamps are more reliable than filament types.
7. Double-pole switches should be used for breaking mains circuits on equipment. Fuses of correct rating should be connected to the equipment side of each switch. (Note C). Always switch off before changing a fuse. The use of AC/DC equipment should be avoided.
8. In metal-enclosed equipment install primary circuit

breakers, such as micro-switches, which operate when the door or lid is opened. Check their operation frequently.
9. Test prods and test lamps should be of the insulated pattern.
10. A rubber mat should be used when the equipment is installed on a floor that is likely to become damp.
11. Switch off before making any adjustments. If adjustments must be made while the equipment is live, use one hand only and keep the other in your pocket. Never attempt two-handed work without switching off first. Use good-quality insulated tools for adjustments.
12. Do not wear headphones while making internal adjustments on live equipment.
13. Ensure that the metal cases of microphones, Morse keys etc are properly connected to the earthed chassis.
14. Do not use meters with metal zero-adjusting screws in high-voltage circuits. Beware of live shafts projecting through panels, particularly when metal grub screws are used in control knobs.
15. Antennas should not, under any circumstances, be connected through a capacitor which may have HT on the other side; a low-resistance DC path to earth should be provided (RF choke).

Note A
Owing to the common use of plastic water main and sections of plastic pipe in effecting repairs, it is no longer safe to assume that a mains water pipe is effectively connected to earth. Steps must be taken, therefore, to ensure that the earth connection is of sufficiently low resistance to provide safety in the event of a fault. Checks should be made whenever repairs are made to the mains water system in the building.

Note B
A 'wandering earth lead' or an 'insulated earthed probe lead' is an insulated lead permanently connected at one end to the chassis of the equipment; at the other end a suitable length of bare wire is provided for touch contacting the high-potential terminals to be discharged.

Note C
Where necessary, surge-proof fuses can be used.

RSGB Mobile Safety Recommendations

1. All equipment should be so constructed and installed that in the event of accident or sudden braking it cannot injure the occupants of the vehicle.
2. Mobile antennas should be soundly constructed, taking into account flexing at speed and possible danger to other vehicles or pedestrians. The maximum height must not exceed 14ft above ground.
3. Wiring should not constitute a hazard, either electrical or mechanical, to driver or passengers.
4. All equipment should be adequately fused and a battery isolation switch is desirable.
5. The transmit/receive switch should be within easy access of the operator and one changeover switch should perform all functions.
6. The microphone should be attached to the vehicle so that it does not impair the vision or movement of the driver.
7. The driver/operator should not use a hand microphone or double headphone.
8. All major adjustments, for example a band change by a driver/operator, should be carried out while the vehicle is stationary.
9. Essential equipment controls should be adequately illuminated during the hours of darkness.
10. All equipment must be switched off when (a) fuelling, (b) in close proximity to petrol tanks and (c) near quarries where charges are detonated electrically.
11. A suitable fire extinguisher should be carried and be readily accessible.

RF and microwave hazards

A great deal of research has been done on the effects of exposure to radio frequencies. It is generally agreed by Western scientists that the only biologically significant property of RF energy is heating, and that this is only a hazard if the heat is not removed quickly by the body's temperature regulating mechanisms. This is the principle of the microwave oven and medical diathermy. RF radiation is therefore totally different in its properties to ionising radiations such as gamma rays and X-rays. The maximum level for continous exposure to RF was set at a power density of 10mW/cm^2 over 30 years ago. This level is still accepted in the UK and the Western world. It has been suggested in Russia and some Eastern countries that there are effects other than thermal ones and that the maximum power density should be 10μW/cm^2. Evidence of this is regarded with suspicion in the West.

Close proximity to a source of RF energy must therefore be avoided. Measurements of power density made in the vicinity of various 300–400W output amateur stations operating on 28MHz and 144MHz and their antennas have given values of less than 1mW/cm^2. Calculation shows that standing 20cm from a vertical $\lambda/4$ antenna fed with 140W at 28MHz is equivalent to being in a 10mW/cm^2 field. This is not, however, a normal situation to be in!

Microwave radiation itself is not more hazardous, but the smaller area over which microwaves are likely to radiate results in a greater power density. Any exposure to RF which results in a sensation of heat is far in excess of 10mW/cm^2 and is therefore very dangerous. The eyes are particularly susceptible to damage in this way.

Situations to be avoided:

1. Do not look down a waveguide unless you know that there is no RF at the other end.
2. Do not work on high-power RF equipment with the covers off. (There is probably high voltage present as well!)
3. Do not use an unscreened dummy load or a small antenna in the shack for test purposes.
4. Do not adjust antennas using full power; use low power only.
5. Do not use a hand-held set without a thick insulating cap on the end of the antenna.
6. Hold the hand-held set so that the antenna is as far as possible from the face.

CHAPTER 13

Tackling the RAE

Many candidates study for the examination at a local technical college or evening institute, but a large number prepare themselves privately either with or without the assistance of a correspondence course. Colleges and institutions offering RAE courses are listed in *Radio Communication* and other radio magazines, usually during July, August and September each year. Early application is advised so that colleges are aware that there will be sufficient students to warrant the course being organised.

If it appears after enquiry that no course is proposed in the reader's locality, it is suggested that the local radio club or RSGB Group be contacted if this has not already been done. Many clubs are affiliated to the Society and their addresses are listed in the RSGB *Amateur Radio Call Book*. Clubs often run a series of talks helpful to prospective amateurs. If, however, enough prospective candidates for the RAE can be found, then an approach to the Principal of the nearest college or institute should be made. Normally 12 candidates will warrant a course being started provided an instructor (preferably a teacher/amateur) and accommodation are available at the college. Courses, one or two evenings each week, usually start in September so make this approach some months beforehand – it might even be possible to find and suggest a suitable instructor as well if the college has not got one! It should be noted that colleges do not normally accept students below the age of 14 and many have a lower age limit of 16.

If no college course is available, do not give up. Many private students are successful every year. Whichever method of preparation is followed, the prospective candidate is strongly urged to contact his or her local club, where practical and willing help is sure to be found. If there appears to be no such club within reach, do not hesitate to contact amateurs in the area (see the *Amateur Radio Call Book*). Most of them will be only too willing to show their radio equipment and give useful help and advice to assist in becoming a radio amateur.

Examination requirements

Don't neglect the study of calculations and manipulation of formulae because the examination questions are multiple-choice. There are many worked examples in Chapter 2 and Appendix 2.

Familiarise yourself with the symbols for electronic components. These are defined in British Standard 3939 and a selection of the common ones are in another document: *Electrical and electronic graphical symbols for schools and colleges* PP7303. The most useful ones for the RAE are given in Appendix 1 of this manual. During studies, note the values of the components in different parts of circuits used for dealing with different frequencies.

In the examination room

It is most important to find out exactly where the examination is to be held and to arrive there 10 minutes or so before the examination commences.

There are a number of formalities to be gone through. The examination regulations have to be read to the candidates by the invigilator. Your lecturer may well be present but he is unlikely to be the invigilator. Your centre number and candidate number will have been notified to you by the college authorities. These numbers together with your name and address and other information have to be written down in several places. This is obviously vitally important!

In the examination you are required to indicate which of the four possible answers you consider to be the correct one by filling in the appropriate box a, b, c or d on the answer sheet with an HB pencil. The latter is most important: an HB pencil is provided. If you wish to change your answer, it is important that you change it in the way shown on the answer sheet.

You will see that you also have to indicate your candidate number and centre number by filling in the numbered boxes at the top right-hand corner of the answer sheet. Thus your answers and identification number can be read by the computer.

Any rough work or calculations can be done on the question book. Note that the question book cannot be taken away: it is a C & G requirement that it be handed in at the end of the examination and returned to C & G.

Do remember the following

The examination questions cover all the objectives. During studies the whole of the syllabus content should be covered. Do not on the one hand imagine that it is so easy that one can hardly fail, or on the other hand feel that the questions are bound to be beyond one's capabilities. Candidates are not competing against others but are trying to reach a certain standard.

If a question is found to be too difficult, leave it and pass on to the next. One can return to it later and make an intelligent guess. The computer will only record the correct answer. It cannot be fooled by filling in all the spaces provided for the answer!

Provided the required standard is reached a pass will be obtained. A great many candidates of all ages and from all walks of life are successful in passing this examination.

Good luck!

Radio circuit symbols

* Indicates preferred symbol

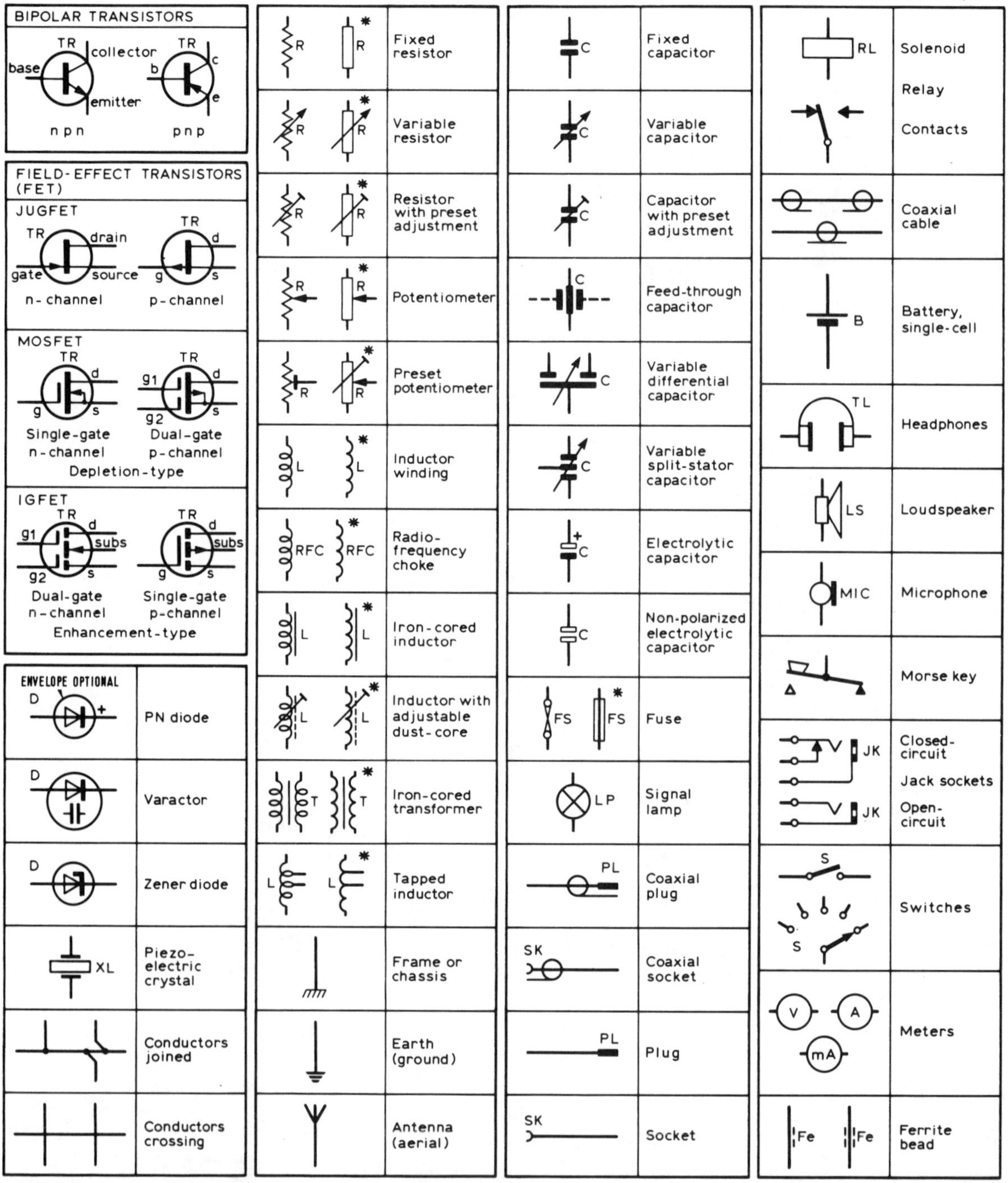

BIPOLAR TRANSISTORS

TR, base, collector, emitter — npn
TR, b, c, e — pnp

FIELD-EFFECT TRANSISTORS (FET)

JUGFET

TR, gate, drain, source — n-channel
TR, g, d, s — p-channel

MOSFET

TR, g, d, s — Single-gate n-channel
TR, g1, g2, d, s — Dual-gate p-channel

Depletion-type

IGFET

TR, g1, g2, d, subs, s — Dual-gate n-channel
TR, g, d, subs, s — Single-gate p-channel

Enhancement-type

Symbol	Name
ENVELOPE OPTIONAL D +	PN diode
D	Varactor
D	Zener diode
XL	Piezo-electric crystal
	Conductors joined
	Conductors crossing

Symbol	Name
R, R *	Fixed resistor
R, R *	Variable resistor
R, R *	Resistor with preset adjustment
R, R *	Potentiometer
R, R *	Preset potentiometer
L, L *	Inductor winding
RFC, RFC *	Radio-frequency choke
L, L *	Iron-cored inductor
L, L *	Inductor with adjustable dust-core
T, T *	Iron-cored transformer
L, L *	Tapped inductor
	Frame or chassis
	Earth (ground)
	Antenna (aerial)

Symbol	Name
C	Fixed capacitor
C	Variable capacitor
C	Capacitor with preset adjustment
C	Feed-through capacitor
C	Variable differential capacitor
C	Variable split-stator capacitor
+ C	Electrolytic capacitor
C	Non-polarized electrolytic capacitor
FS, FS *	Fuse
LP	Signal lamp
PL	Coaxial plug
SK	Coaxial socket
PL	Plug
SK	Socket

Symbol	Name
RL	Solenoid Relay Contacts
	Coaxial cable
B	Battery, single-cell
TL	Headphones
LS	Loudspeaker
MIC	Microphone
	Morse key
JK JK	Closed-circuit Jack sockets Open-circuit
S S	Switches
V, A, mA	Meters
Fe, Fe	Ferrite bead

APPENDIX 2

Mathematics for the RAE

The basic mathematical processes are: addition, subtraction, multiplication and division. As long as only 'whole numbers' are involved, such sums are simple.

However, very often we must consider quantities which are which are less than one (unity), for instance $^1/_2$, $^1/_3$, $^1/_8$ etc. Here $^1/_8$ means one-eighth part of the whole and so on.

$^1/_8$ is called a vulgar fraction and has two parts: the '8' (the bottom part) is called the 'denominator' and the '1' (the top part) is called the 'numerator'. The magnitude of a fraction is not changed if we multiply top and bottom by the same number, ie

$$\frac{3}{16} \times \frac{2}{2}$$

As the '2' is on the top and bottom we can 'cancel' it thus:

$$\frac{3}{16} \times \frac{\not{2}}{\not{2}} = \frac{3}{16}$$

A fraction should always be cancelled down to its simplest form:

$$\frac{12}{16} = \frac{3 \times \not{2}}{4 \times \not{2}} = \frac{3}{4}$$

Here top and bottom have been divided by 4.

Fractions can be

(a) Multiplied

$$\frac{1}{2} \times \frac{3}{4} \times \frac{5}{8} = \frac{1 \times 3 \times 5}{2 \times 4 \times 8} = \frac{15}{64}$$

(b) Divided

$$\frac{3}{4} \div \frac{1}{2}$$

Dividing by $^1/_2$ is the same as multiplying by $^2/_1$, ie

$$\frac{3}{4} \div \frac{1}{2} = \frac{3}{4} \times \frac{2}{1} = \frac{6}{4} = 1\frac{2}{4} = 1\frac{1}{2}$$

In other words, dividing by a fraction is the same as multiplying by that fraction 'upside down'. Another example is:

$$\frac{7}{8} \div \frac{3}{4} = \frac{7}{8} \times \frac{4}{3} = \frac{7}{2} \times \frac{1}{3} = \frac{7}{6} = 1\frac{1}{6}$$

Here we divide top and bottom by 4.

(c) Added

$$\frac{2}{3} + \frac{2}{3} = \frac{2 + 2}{3} = \frac{4}{3}$$

If the denominators are different, we must make them the same, ie 'bring them to a common denominator' and normally the lowest common denominator is used. For example

$$\frac{2}{3} + \frac{5}{6} = \frac{4}{6} + \frac{5}{6} = \frac{9}{6} = \frac{3}{2} = 1\frac{1}{2}$$

Here we have multiplied top and bottom of $^2/_3$ by 2, making it $^4/_6$. Hence we can add it to $^5/_6$, making $^9/_6$, which is then simplified to $1^1/_2$. Another example is

$$\frac{1}{3} + \frac{5}{6} + \frac{7}{8} = \frac{8}{24} + \frac{20}{24} + \frac{21}{24} = \frac{8 + 20 + 21}{24}$$

$$= \frac{49}{24} = 2\frac{1}{24}$$

It is generally preferable to divide out fractions greater than one, as we have done above.

(d) Subtracted

Exactly the same rules apply to the subtraction of fractions.

We can also express parts of the whole as 'decimals' or $^1/_{10}$ parts, written as 0.1, 0.2, 0.3 etc (these are equivalent to $^1/_{10}$, $^2/_{10}$, $^3/_{10}$ etc). The 'full stop' is known as the 'decimal point'. In a decimal, the 'nought' before the decimal point should never be omitted.

The denominator of any fraction can be divided into the numerator to give a decimal, eg

$^1/_8$ = 0.125

$^3/_8$ = 0.375

The more common fractions and decimal equivalents should be memorised, eg

$^1/_{10} = 0.1$	$^1/_8 = 0.125$
$^2/_{10} = {}^1/_5 = 0.2$	$^2/_8 = {}^1/_4 = 0.25$
$^3/_{10} = 0.3$	$^3/_8 = 0.375$
$^4/_{10} = {}^2/_5 = 0.4$ etc	$^4/_8 = {}^1/_2 = 0.5$ etc

Numbers can be expressed to 'so many significant figures' or 'so many decimal places'.

Thus 12345 is a number to five significant figures
1234 is a number to four significant figures
123 is a number to three significant figures

Note also that 1.23 is a number to three significant figures (the decimal point is ignored).

12.345 is a number to three decimal places
12.34 is a number to two decimal places
12.3 is a number to one decimal place

Decimals may be 'rounded off'; that means

3.3267 to three decimal places is 3.327
(the 7 is greater than 5, so 6 becomes 7)
3.327 to two decimal places is 3.33
(the 7 is greater than 5, so 2 becomes 3)
3.33 to one decimal place is 3.3
(the 3 is less than 5, so is ignored)

Powers of numbers

When, say, two of a certain number are multiplied together, that number is said to be 'raised to the power 2'. Thus $2 \times 2 = 4$ means that 2 raised to the power 2 is 4. In this case we would say 2 'squared' is 4 and write it as $2^2 = 4$. The 'little 2 up in the air' is called an 'index'. Similarly, $2 \times 2 \times 2 = 8$ means that 2 raised to the power 3 is 8, or 2 'cubed' is 8, written as $2^3 = 8$. Also $2 \times 2 \times 2 \times 2 = 16$. Here we have no alternative but to say 2 'to the power 4' = 16, or $2^4 = 16$.

The use of indices or the index notation is a very convenient way of expressing the large numbers which often occur in radio calculations, eg

$$100 = 10 \times 10 = 10^2$$

$$10{,}000 = 10 \times 10 \times 10 \times 10 = 10^4$$

$$1{,}000{,}000 = 10 \times 10 \times 10 \times 10 \times 10 \times 10 = 10^6$$

Note that $10 = 10^1$ (the index here is taken for granted). Similarly

$$\frac{1}{100} = \frac{1}{10 \times 10} = \frac{1}{10^2} \text{ (written as } 10^{-2}\text{)}$$

$$\frac{1}{10{,}000} = \frac{1}{10 \times 10 \times 10 \times 10} = \frac{1}{10^4} \text{ (written as } 10^{-4}\text{)}$$

$$\frac{1}{1{,}000{,}000} = \frac{1}{10 \times 10 \times 10 \times 10 \times 10 \times 10} = \frac{1}{10^6} \text{ (written as } 10^{-6}\text{)}$$

Numbers expressed in the index notation are multiplied and divided by adding and subtracting respectively the indices.

$$10^2 \times 10^3 = 10^{2+3} = 10^5$$

$$10^4 \div 10^2 = 10^{4-2} = 10^2$$

$$\frac{10^5 \times 10^7 \times 10^{-2}}{10^3 \times 10 \times 10^{-3}} = \frac{10^{5+7-2}}{10^{3+1-3}} = \frac{10^{10}}{10^1} = 10^9$$

We can do this as long as the 'base' is the same in each case. In the above examples, the 'base' is 10. For example, $10^2 \times 2^2 = 100 \times 4 = 400$, which is neither 10^4 or 2^4!

Roots of numbers

The root of a number is that number which, when multiplied by itself so many times, equals the given number; the 'square' root of 4 is 2, ie $2 \times 2 = 4$, and this is written $\sqrt[2]{4} = 2$.

Similarly the 'cube' root of 8 is 2, ie $2 \times 2 \times 2 = 8$, and $\sqrt[4]{16} = 2$ etc. Note the little 2 in the sign for square root is normally omitted so that $\sqrt{}$ signifies the square root.

Numbers like 4, 16 and 25 are called 'perfect' squares because their square roots are whole numbers, thus

$$\sqrt{49} = 7 \qquad \sqrt{121} = 11 \quad \text{etc}$$

The following should be memorised as they can often be very useful.

$$\sqrt{2} = 1.41 \quad \sqrt{3} = 1.73 \quad \sqrt{5} = 2.24 \quad \sqrt{10} = 3.162$$

For example

$$\sqrt{200} = \sqrt{2 \times 100} = \sqrt{2} \times \sqrt{100} = 1.41 \times 10 = 14.1$$

$$\sqrt{192} = \sqrt{3 \times 64} = \sqrt{3} \times \sqrt{64} = 1.73 \times 8 = 13.8$$

It is always worth checking to see if the number left after dividing by 2, 3 or 5 is a perfect square.

The square root of a number expressed in the index notation is found by dividing the index by 2, thus $\sqrt{10^6} = 10^3$ and $\sqrt{10^{12}} = 10^6$, and so on. Similarly $\sqrt{10^{-6}} = 10^{-3}$ etc. Should the index be an odd number, it must be made into an even number as follows.

$$\sqrt{10^{-15}} = \sqrt{10 \times 10^{-16}}$$

$$= \sqrt{10} \times 10^{-8}$$

$$= 3.162 \times 10^{-8}$$

The constant term 'π' occurs in many calculations; 'π' has great significance in mathematics and is the ratio of the circumference to the diameter of a circle. π can be taken to be 3.14 or 22/7. The error in taking π^2 as 10 is less than 1.5 per cent and is acceptable here. $1/\pi$ can be taken as 0.32 and $1/2\pi$ as 0.16 (the error in calling this $^1/_6$ is really somewhat too high). $1/2\pi = 0.16$ is particularly useful.

Typical calculations

We will now apply these rules to the solution of problems likely to be met in radio work as a lead-in to some typical numerical multiple-choice questions.

Answers to three significant figures as given by a slide rule or four-figure logarithm tables are satisfactory for most radio purposes and the eight figures given by the electronic calculator should certainly be rounded off.

The most important aspect is to remember that the units met with are most likely to be the practical ones such as microfarads, picofarads, milliamperes, millihenrys etc. These must be converted into the basic units of farads, amperes and henrys before substituting them into the appropriate formula. This involves multiplying or dividing by 1000 (10^3), 1,000,000 (10^6) and so on. Therefore the important thing is to get the decimal point in the right place or the right number of noughts in the answer. The commonest conversions are as follows:

There are 10^6 microfarads in 1 farad
hence $8\mu F = 8 \times 10^{-6}$ farads
There are 10^{12} picofarads in one farad
hence $22pF = 22 \times 10^{-12}$ farads

(The use of 'nano' or 10^{-9} is now fairly common; there are 10^9 nanofarads in 1 farad so $1nF = 1 \times 10^{-9}$ farads, but such a capacitor may well be marked '1000pF'.) Similarly other conversions are

$50\mu H = 50 \times 10^{-6}$ henrys
$3mH = 3 \times 10^{-3}$ henrys
$45mA = 45 \times 10^{-3}$ amperes
$10\mu A = 10 \times 10^{-6}$ amperes

Problem 1

What value of resistor is required to drop 150V when the current flowing through it is 25mA?

This involves Ohm's Law which can be expressed in symbols in three ways:

$$R = \frac{V}{I} \qquad I = \frac{V}{R} \qquad V = I \times R$$

where R is in ohms, V in volts and I in amperes. Clearly the first, $R = V/I$, is needed. First of all, we must express the current (25mA) in amperes.

$$25mA = \frac{25}{1000} \text{ A (or } 25 \times 10^{-3}\text{A)}$$

Substituting values for V and I

$$R = \frac{V}{I}$$

$$= 150 \times \frac{1000}{25}$$

(we are dividing by $^{25}/_{1000}$, ie multiplying by $^{1000}/_{25}$)

hence $$R = \frac{150 \times 1000}{25}$$

25 'goes into' 150 six times, so

$$R = 6 \times 1000$$

$$= 6000\Omega$$

Problem 2

What power is being dissipated by the resistor in Problem 1?

The power dissipated in the resistor is power (watts) = V (volts) $\times$ I (amps). By Ohm's Law, power can be expressed in two other forms.

$$W = \frac{V^2}{R} \text{ and } W = I^2R$$

Because we know V, I and R we can use any of the above relationships, say

$$W = \frac{V^2}{R}$$

$$W = \frac{150 \times 150}{6000}$$

Two 'noughts' on the top and bottom can be cancelled, leaving

$$= \frac{15 \times 15}{60}$$

Cancelling 15 into 60 leaves

$$= \frac{15}{4} = 3\tfrac{3}{4}\text{W}$$

The other two forms will, of course, give the same answer – try them!

Problem 3

Resistors of 12Ω, 15Ω and 20Ω are in parallel. What is the effective resistance?

$$\frac{1}{R} = \frac{1}{R_1} + \frac{1}{R_2} + \frac{1}{R_3}$$

$$= \frac{1}{12} + \frac{1}{15} + \frac{1}{20}$$

60 is the lowest common denominator of 12, 15 and 20, so

$$\frac{1}{R} = \frac{5}{60} + \frac{4}{60} + \frac{3}{60}$$

$$= \frac{5 + 4 + 3}{60}$$

$$= \frac{12}{60}$$

This is a simple equation in R, and the first step in solving it is to 'cross-multiply'. It may be shown that the denominator of one side multiplied by the numerator of the other side is equal to the numerator of the first side multiplied by the denominator of the other side, thus

$$R \times 12 = 1 \times 60$$

Hence, dividing each side by 12

$$R = \frac{60}{12}$$

$$R = 5\Omega$$

Problem 4

Capacitors of 330pF, 680pF and 0.001µF are in parallel. What is the effective capacitance?

The first step is to express all the capacitors in the *same* units which can be either picofarads or microfarads.

$$0.001\mu F = 0.001 \times 1{,}000{,}000pF$$

(there are 1,000,000pF in 1µF) and hence

$$0.001\mu F = 1000pF$$

Effective capacitance is therefore

$$330pF + 680pF + 1000pF = 2010pF$$

Problem 5

What is the reactance of a 30H smoothing choke at a frequency of 100Hz?

$$X_L = 2\pi fL$$

$$X_L = 2\pi \times 100 \times 30 \quad \text{ohms}$$

$$= 6000\pi \text{ ohms}$$

We take π to be 3.14 so

$$X_L = 6000 \times 3.14$$

$$= 18{,}840\Omega$$

Problem 6

What is the reactance of a 100pF capacitor at a frequency of 20MHz?

$$X_C = \frac{1}{2\pi fC}$$

(X_C is in ohms when f is in hertz, L in henrys and C in farads.)

$$f = 20MHz = 20 \times 10^6 Hz = 2 \times 10^7 Hz$$

$$C = 100pF = 100 \times 10^{-12}F = 10^{-10}F$$

(It is much more convenient here to use the index notation.)
Hence

$$X_C = \frac{1}{2\pi \times 2 \times 10^7 \times 10^{-10}} \quad \text{ohms}$$

$$= \frac{1}{2\pi} \times \frac{1}{2 \times 10^{-3}}$$

Note that we have kept $^1/_{2\pi}$ intact because $^1/_{2\pi} = 0.16$, thus

$$X_C = 0.16 \times \frac{1}{2 \times 10^{-3}}$$

$$= \frac{0.16 \times 1000}{2}$$

$$= 80\Omega$$

Problem 7

What is the impedance (Z) of an inductance which has a resistance (R) of 4Ω and a reactance (X) of 3Ω?

$$Z = \sqrt{(R^2 + X^2)}$$

$$= \sqrt{(4^2 + 3^2)}$$

$$= \sqrt{16 + 9}$$

$$= \sqrt{25}$$

$$= 5\Omega$$

Problem 8

At what frequency do a capacitor of 100pF and an inductance of 100µF resonate?

At resonance

$$2\pi fL = \frac{1}{2\pi fC}$$

hence

$$f = \frac{1}{2\pi\sqrt{LC}}$$

(f is in hertz, L is in henrys, C is in farads.)

$$100\mu H = 100 \times 10^{-6}H$$

$$100pF = 100 \times 10^{-12}F$$

$$f = \frac{1}{2\pi\sqrt{LC}}$$

$$= \frac{1}{2\pi\sqrt{100 \times 10^{-6} \times 100 \times 10^{-12}}}$$

$$= \frac{1}{2\pi\sqrt{10^2 \times 10^{-6} \times 10^2 \times 10^{-12}}}$$

$$= \frac{1}{2\pi\sqrt{10^{-14}}}$$

$$= \frac{1}{2\pi \times 10^{-7}}$$

$$= \frac{1}{2\pi} \times 10^7$$

$$= 0.16 \times 10^7$$

$$= 1.6 \times 10^6\text{Hz}$$

$$= 1.6\text{MHz}$$

Numerical multiple-choice questions in the RAE

The numerical multiple-choice questions set in the RAE involve quite simple calculations in order to decide which of the four answers given is correct. The questions are likely to be similar to the problems just worked through and generally the answer comes out without the need for any aid to calculation. As in solving the previous problems, the most important thing is to 'get the units right'. The way to solve these questions should be clear from the following worked examples.

Question 1

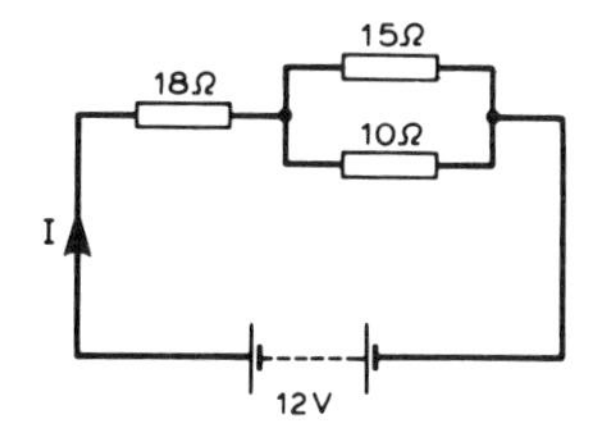

The current *I* is
(a) 0.25A.
(b) 0.43A.
(c) 0.5A.
(d) 0.67A.

The effective resistance of the two resistors in parallel is

$$R_{\text{eff}} = \frac{15 \times 10}{25} = 6\Omega$$

The effective resistance of the whole circuit is

$$R_{\text{eff}} = 18 + 6 = 24\Omega$$

$$I = \frac{12}{24} = 0.5\text{A}$$

Answer (c) is therefore correct.

Question 2

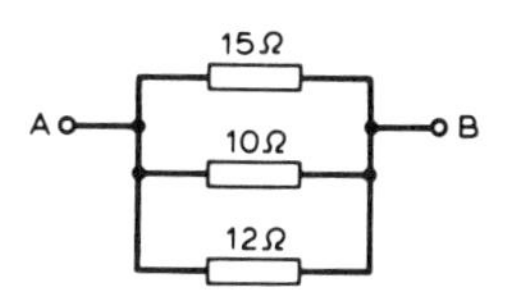

The effective resistance between points A and B is
(a) 4Ω.
(b) 6Ω.
(c) 17Ω.
(d) 37Ω.

The effective resistance must have a value less than the value of the smallest resistor, so neither answers (c) nor (d) are correct. Take the top two resistors and apply the formula

$$R_{\text{eff}} = \frac{R_1 \times R_2}{R_1 + R_2} = \frac{15 \times 10}{25} = 6\Omega$$

Again apply the formula to include the 12Ω resistor.

$$R_{\text{eff}} = \frac{6 \times 12}{18} = \frac{72}{18} = 4\Omega$$

Answer (a) is therefore correct.

Question 3

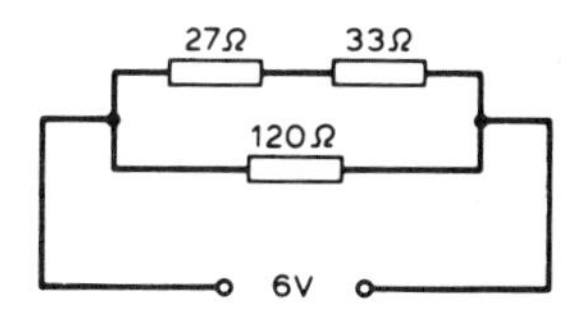

The current flowing through the 27Ω resistor has a value of
(a) 27mA.
(b) 33mA.
(c) 60mA.
(d) 100mA.

The current flowing through the 120Ω resistor has no bearing on the answer. The current through the 27Ω resistor will be the same as that through the 33Ω resistor. The current through the two resistors in series

$$= \frac{6}{27 + 33} = \frac{6}{60} = \frac{1}{10}\ \text{A}$$

The correct answer is (d).

Question 4

The voltage across the 33Ω resistor in the previous question is
(a) 0.6V.
(b) 1.2V.
(c) 3.3V.
(d) 4.5V.

$$V = I \times R = \frac{1}{10} \times 33 = 3.3\text{V}$$

The correct answer is (c).

Question 5

A $\lambda/2$ dipole has a length of just under 7.5m. It will be resonant at a frequency of approximately
(a) 15MHz.
(b) 20MHz.
(c) 25MHz.
(d) 30MHz.

$$\lambda = 15\text{m} \quad f = \frac{c}{\lambda} = \frac{300 \times 10^6}{15} = 20 \times 10^6\text{Hz} = 20\text{MHz}$$

Therefore (b) is the correct answer.

Question 6

An oscilloscope shows the peak-to-peak voltage of a sine wave to be 100V. The RMS value is
(a) 27.28V.
(b) 35.35V.
(c) 50V.
(d) 70.7V.

$$V_{rms} = V_{peak} \times 0.707$$
$$= 50 \times 0.707 = 35.35\text{V}$$

The correct answer is (b).

Question 7

The internal capacitance between the base and emitter of a transistor is 2pF. The reactance at a frequency of 500MHz will be approximately
(a) 16Ω.
(b) 160Ω.
(c) 1.6kΩ.
(d) 16kΩ.

$$X_C = \frac{1}{\omega C} = \frac{1}{2\pi \times 500 \times 10^6 \times 2 \times 10^{-12}} = \frac{1}{2\pi \times 10^{-3}}$$
$$= 0.16 \times 10^3 = 160\Omega$$

The correct answer is (b).

Question 8

A loudspeaker speech coil has a resistance of 3Ω. If the voltage across it is 3V, then the power in the speech coil is
(a) 1W
(b) 3W.
(c) 6W.
(d) 9W.

$$P = \frac{V^2}{R} \text{ watts} \qquad P = \frac{3^2}{3} = 3\text{W}$$

Answer (b) is therefore correct.

Question 9

A smoothing choke has an inductance of 0.2H. Its reactance at a frequency of 100Hz is approximately
(a) 40Ω.
(b) 125Ω.
(c) 400Ω.
(d) 1250Ω.

$X_L = 2\pi fL = 2\pi \times 100 \times 0.2 = 40\pi$ or about 125Ω. Hence (b) is the correct answer.

Question 10

A coil has a reactance of 1000Ω and a resistance of 10Ω. Its approximate impedance is
(a) 990Ω.
(b) 1000Ω.
(c) 1100Ω.
(d) 10kΩ.

$$Z = \sqrt{R^2 + X_L^2} = \sqrt{100 + 10^6} = \sqrt{1{,}000{,}100} \approx 1000\Omega$$

The effect of the resistance is so small that it can be neglected, so (b) is the correct answer.

Question 11

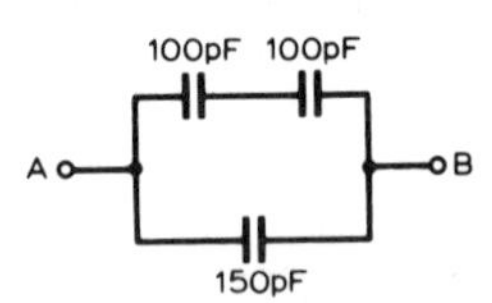

The capacitance measured between terminals A and B will be
(a) 37.5pF.
(b) 50pF.
(c) 200pF.
(d) 350pF.

The capacitance must be greater than 150pF. The two 100pF capacitors in series have an effective capacitance of 50pF. Therefore the answer is 150 + 50 = 200pF, ie answer (c).

Question 12

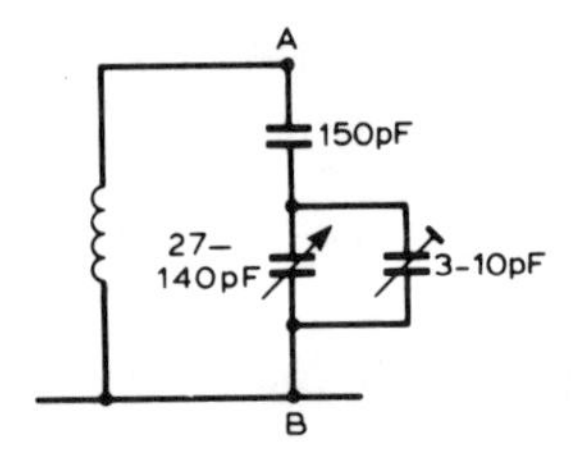

When the variable capacitor and the trimmer capacitor of a local oscillator tuned circuit are adjusted to their maximum values, the effective capacitance between points A and B will be

(a) 50pF.
(b) 75pF.
(c) 200pF.
(d) 300pF.

The tuning and trimmer capacitors will have an effective capacitance of 140 + 10 = 150pF. Therefore the capacitance between A and B will be 75pF, and (b) is the correct answer.

Question 13

When the variable capacitor and the trimmer are set at minimum, the effective capacitance between points A and B will be
(a) 25pF.
(b) 75pF.
(c) 120pF.
(d) 180pF.

Using the same calculations, (a) is the correct answer.

Question 14

The peak-to-peak value of a sine wave having an RMS value of 14.1V is approximately
(a) 20V.
(b) 28.2V
(c) 40V.
(d) 56.4V.

The peak value of the positive half-cycle is $14.1 \times \sqrt{2} = 20$V. Therefore the peak-to-peak voltage = $2 \times 20 = 40$V.

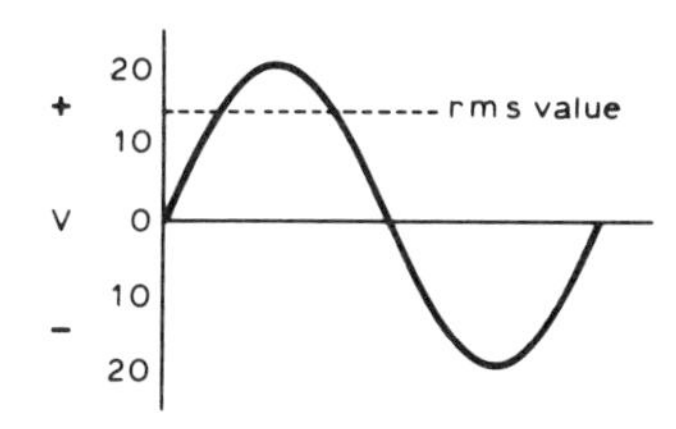

(c) is the correct answer.

Question 15

A quarter-wave antenna is resonant at 10MHz. Its approximate length will be
(a) 7.5m.
(b) 15m.
(c) 20m.
(d) 30m.

$$\lambda = \frac{C}{f} = \frac{300 \times 10^6}{10 \times 10^6} = 30\text{m}$$

$$\lambda/4 = 7.5\text{m}$$

(a) is the correct answer.

Summary of formulae

Ohm's Law $R = \frac{V}{I}$ $V = IR$ $I = \frac{V}{R}$

Power $W = V \times I$ $W = I^2R$ $W = \frac{V^2}{R}$

Reactance $X_L = 2\pi fL$

$$X_C = \frac{1}{2\pi fC}$$

Resonance $f_r = \frac{1}{2\pi\sqrt{LC}}$

Resistors (series) $R = R_1 + R_2 + R_3 + \ldots$

Resistors (parallel) $\frac{1}{R} = \frac{1}{R_1} + \frac{1}{R_2} + \frac{1}{R_3} + \ldots$

Capacitors (series) $\frac{1}{C} = \frac{1}{C_1} + \frac{1}{C_2} + \frac{1}{C_3} + \ldots$

Capacitors (parallel) $C = C_1 + C_2 + C_3 + \ldots$

Wavelength (metres) $= \frac{300}{f(\text{MHz})}$

For a sine wave, RMS value = 0.707 × peak value

$$Q = \frac{\omega L}{R}$$

$$R_D = \frac{L}{CR}$$ and since $Q = \frac{\omega L}{R}$

$$R_D = \frac{Q}{\omega C}$$

$$e = \frac{7.02\sqrt{\text{ERP}}}{d}$$

where *e* is the field strength (peak) in volts per metre, *d* is the distance in metres from the transmitter, and the ERP is measured in watts.

APPENDIX 3

Radio Amateurs' Examination objectives and syllabus

The examination objectives describe in general terms the nature of the examination questions, while the syllabus states the subject matter to which they relate.

PAPER 1. LICENSING CONDITIONS AND TRANSMITTER INTERFERENCE AND ELECTROMAGNETIC COMPATIBILITY (EMC)

1. Licensing conditions

Examination objectives

1. Name the types and state the purposes of Amateur Licences available. State the qualifications necessary to hold a licence.
2. State the conditions of the Amateur Licences (A) and (B) and show a knowledge of the notes appended to them.

Syllabus

1. Types of licence available and the qualifications necessary.
2. Conditions (terms and limitations) laid down in the Amateur Licences (A) and (B) including the notes appended and an understanding of the schedule of frequency, bands, power and types of transmission.

2. Transmitter interference

Examination objectives

1. Describe the consequences of poor frequency stability.
2. For spurious emissions
   (i) describe their causes;
   (ii) describe methods, appropriate to the Amateur Service, of detecting and recognising their presence;
   (iii) describe in practical terms the measures which should be taken in both the design and construction of transmitters and the use of filters to minimise them.
3. Demonstrate knowledge of frequency-checking equipment.

Syllabus

2.1. Frequency stability; consequences of poor frequency stability; risks of interference, out-of-band radiation.

2.2. Spurious emissions, causes and methods of prevention; harmonics of the radiated frequency, direct radiation from frequency-determining stages (including synthesisers) of a transmitter, parasitic oscillations, excessive sidebands due to overmodulation. Excessive deviation of FM transmitters. Key clicks, methods of suppression.

2.3. Frequency checking.

3. Electromagnetic Compatibility (EMC)

Examination objectives

1. Demonstrate knowledge of EMC in respect of problems that can be caused when an amateur station operates in close proximity to other electronic equipment.
2. Describe and explain interference sources.
3. Describe and explain the various paths by which interfering signals can arrive at domestic TV, radio and audio equipment.
4. Describe and explain methods of improving the immunity of affected equipment.
5. Describe and explain methods of improving the amateur station design to reduce EMC problems.
6. Explain the steps to be taken when investigating EMC problems with a neighbour's equipment.

Syllabus

3.1. EMC: the ability of a device, equipment or system to function satisfactorily in its electromagnetic environment without introducing intolerable electromagnetic disturbances to anything in that environment. EMC problems that are likely to occur when an amateur station operates in close proximity to other electronic equipment.

3.2. Equipment used in an amateur station that is capable of generating broad-band and narrow-band interference.

3.3. Interfering signal paths: RF, IF, audio and mains borne.

3.4. Methods of improving the immunity of affected equipment:
   (i) Use of toroidal chokes and filters (mains, high pass, low pass, bandpass, notch or bandstop).

(ii) Characteristics of filters, bandwidth, insertion loss and impedance.
(iii) Screening, lead lengths, and fitting ferrite rings and beads and bypass capacitors.

3.5. Improving station design:
(i) RF grounding.
(ii) Station mains filtering.
(iii) Screening.
(iv) Monitoring output power and calculation of field strengths.
(v) Use of minimum transmitted power.
(vi) Monitoring output transmissions for spurious and harmonic levels, including key-clicks.
(vii) Location of antennas and masts.
(viii) Type and size of antennas.
(ix) Use of screened feeder cables, balanced lines and baluns.

3.6. Method of approach and basic checks required when investigating EMC problems with a neighbour's equipment.

PAPER 2. OPERATING PROCEDURES, PRACTICES AND THEORY

1. Operating procedures

Examination objectives

1. Describe calling procedures in telegraphy, including the use of Q-codes and telephony.
2. Demonstrate knowledge of maintaining a log.
3. For satellites and repeaters
   (i) explain why they are used in the Amateur Service;
   (ii) describe the method of accessing a repeater.
4. Explain the reasons for using Q-codes and other abbreviations.
5. State the reasons for band planning.
6. Demonstrate knowledge of the phonetic alphabet and explain why it is used.
7. Describe the need for safety precautions in an amateur station.

Syllabus

1.1. Calling procedures in telegraphy and telephony: general calls to all stations and calls to specific stations.
1.2. Log-keeping: Licence requirements.
1.3. Use of satellites and repeaters; accessing a repeater.
1.4. Reasons for and use of Q-codes and other abbreviations.
1.5. Band planning: purposes and advantages.
1.6. The phonetic alphabet: reasons for its use.
1.7. Safety in the amateur station:
(i) Precautions recommended by the Radio Society of Great Britain.
(ii) Reasons why equipment to be repaired should be disconnected from the mains supply and capacitors discharged.

2. Electrical theory

Examination objectives

1. For DC circuits
   (i) state Ohm's Law and use it to solve simple problems;
   (ii) calculate total current in series and parallel circuits;
   (iii) calculate power in a DC circuit;
   (iv) calculate the effective resistance of resistors in series and parallel circuits.
2. For AC circuits
   (i) state the terms which define the sine wave;
   (ii) calculate the effective capacitance in series and parallel circuits;
   (iii) the effective inductance in series circuits;
   (iv) explain what is meant by inductive reactance, capacitive reactance and impedance;
   (v) explain the effects of inductive reactance, capacitive reactance and impedance;
   (vi) solve simple problems.
3. For tuned circuits
   (i) state the characteristics and calculate the resonant frequency of series and parallel tuned circuits;
   (ii) explain the magnification of current and voltage at resonance.
4. Explain how a transformer functions and describe its uses.
5. Explain the use of the decibel to express ratios and levels of power and voltage.

Syllabus

2.1. (i) The meaning of basic electrical terms: voltage, current, conductor, insulator, resistance.
2.1. (ii) The relationship between voltage, current and power in the DC circuit.
2.2. (i) The sine wave: definition of amplitude, frequency and period, peak, peak-to-peak, instantaneous, average and RMS values. Simple explanation of the terms: phase angle, phase difference, phase lag and lead.
2.2. (ii) Inductance and capacitance; units, inductive and capacitive reactance. Reactance, impedance and power in the AC circuit.
2.3. Series- and parallel-tuned circuits, resonance, impedance, dynamic resistance, calculation of resonant frequency amplification of current and voltage at resonance. Q (magnification) factor.
2.4. Function and uses of the transformer.
2.5. Simple explanation of how the decibel notation is used to express ratios of power and voltage and how it may also be used to define power levels.

3. Solid-State Devices

Examination objectives

1. Explain in simple terms
   (i) the operation of the junction diode and its use as a rectifier, voltage regulator and variable capacitor;

(ii) operation of NPN and PNP transistors and their use in the three common configurations, emphasising biasing and input and output impedance;
(iii) the field-effect transistor: how its characteristics differ from NPN and PNP transistors;
(iv) the integrated circuit.

2. Explain the use of the transistor as a switch.
3. Describe the application of solid-state devices in receivers and transmitters.
4. Describe the principles of operation of typical power supplies for solid-state equipment.

Syllabus

3.1. Characteristics of junction diodes, NPN, PNP and field-effect transistors.

3.2. The common transistor circuit configurations, emphasising the biasing arrangements and conditions, and input and output impedances.

3.3. Use of solid-state devices as
(i) audio and radio frequency amplifiers;
(ii) oscillators;
(iii) frequency multipliers;
(iv) mixers;
(v) demodulators;
(vi) switches.

3.4. Rectification, smoothing and voltage stabilisation arrangements in low-voltage power supplies.

4. Receivers

Examination objectives

For the superheterodyne receiver, explain
(i) the principle of operation;
(ii) choice of intermediate frequencies; adjacent-channel and image-frequency (second-channel) interference;
(iii) demodulation, reception of morse and telephony (SSB and FM);
(iv) simple automatic gain control (AGC).

Syllabus

4.1. The superheterodyne principle of reception. Principles of reception of morse and telephony (SSB and FM) in terms of radio-frequency amplification; frequency changing; demodulation or detection; audio amplification.

4.2. Advantages and disadvantages of high and low intermediate frequencies; adjacent-channel and image-frequency (second-channel) interference and their avoidance.

4.3. Typical receivers; use of a beat-frequency oscillator. Characteristics of a single-sideband signal and the purpose of a carrier insertion oscillator. Characteristics of an FM signal and the purpose of a ratio detector.

4.4. Reason for automatic gain control; explanation of simple RF-derived system.

5. Transmitters

Examination objectives

1. For oscillators
(i) describe their construction;
(ii) state the factors affecting their stability.
2. Describe the function of the stages in transmitters: oscillator, buffer, frequency multiplier, power amplifier.
3. Explain the procedure for the adjustment and tuning of transmitters, including the use of a dummy load.
4. For modulation and types of emission (SSB and FM)
(i) describe and explain the principles of modulation;
(ii) describe the procedure for adjusting the level of modulation;
(iii) state the relative advantages of each mode.
5. The use of a valve as a power amplifier.

Syllabus

5.1. Oscillators used in transmitters; stability of variable-frequency and crystal-controlled oscillators; their construction and factors affecting stability. Synthesizers, advantages and disadvantages: purpose of each stage with block diagram.

5.2. Transmitter stages: function of frequency changers, frequency multipliers, high- and low-power amplifiers (including linear types).

5.3. Transmitter tuning and adjustment.

5.4. Methods of modulation and types of emission (SSB and FM). Adjustment of modulation level. Relative advantages of morse and telephony (SSB and FM).

5.5 Valves. Their application as power amplifiers, advantages and disadvantages.

6. Propagation and antennas

Examination objectives

1. Define the basic terms.
2. For electromagnetic waves
(i) explain their generation;
(ii) state the relationship between electric and magnetic components.
3. For the ionosphere, troposphere and upper atmosphere
(i) describe in simple terms the structure of the ionosphere;
(ii) explain in simple terms the refracting and reflecting properties of the ionosphere and the troposphere;
(iii) explain the factors which affect the ionisation of the upper atmosphere;
(iv) state the effect of variations of ionisation of the upper atmosphere on the propagation of electromagnetic waves.
4. Describe in simple terms ionospheric (or sky-wave), ground-wave and tropospheric propagation.
5. Explain fading and forms of fade-out.

6. For radio waves
   (i) state their velocity in free space;
   (ii) state the relationship between velocity, frequency and wavelength;
   (iii) calculate frequency and wavelength from given data.
7. For antennas and transmission lines
   (i) describe and explain their operation and construction;
   (ii) explain the principles of coupling and matching antennas to transmitter and receivers;
   (iii) identify from diagrams typical coupling and matching arrangements: antenna tuning units (ATU); simple types based on the parallel-tuned circuit; pi networks
8. Describe balanced and unbalanced feeders and explain the principles of propagation of radio waves along transmission lines; the causes and effects of standing waves.

Syllabus

6.1. Explanation of basic terms: ionosphere, troposphere, atmosphere, field strength, polarisation, maximum usable frequency, critical frequency, skip distance, sunspot cycle.

6.2. Generation of electromagnetic waves: relationship between electric and magnetic components.

6.3. Structure of the ionosphere. Refracting and reflecting properties of the ionosphere and troposphere. Effect of sunspot cycle, winter and summer seasons, and day and night on the ionisation of the upper atmosphere: effect of variations of ionisation on the propagation of electromagnetic waves.

6.4. Ground-wave, ionospheric and tropospheric propagation.

6.5. Fade-out and types of fading, including selective fading. Polarisation, absorption and skip.

6.6. Velocity of radio waves in free space; relationship between velocity of propagation, frequency and wavelength; calculation of frequency and wavelength.

6.7. Receiving and transmitting antennas. Operation and construction of typical antennas, including multiband and directional types; end fed, dipole, ground plane, Yagi and quad. Their directional properties. Coupling and matching.

6.8. Transmission lines; balanced and unbalanced feeders; elementary principles of propagation of radio waves along transmission lines; velocity ratio, standing waves.

7. Measurements

Examination objectives

1. For the measurement of AC, DC and RF voltages and currents
   (i) state the types of instrument in common use: analogue and digital multimeters, oscilloscopes;
   (ii) explain how errors can be caused by the effects of the instrument on the circuit.
2. Explain the method of measurement of DC power input and RF power output of power amplifiers. Explain how power may be expressed in dBW.
3. For frequency-measuring instruments (absorption wavemeters, crystal calibrators and digital frequency meters)
   (i) state the purpose for which they are used;
   (ii) state their accuracy;
   (iii) describe in detail their use at an amateur transmitting station.
4. Describe the construction of dummy loads and explain their use.
5. Explain the purpose and method of using a standing-wave ratio meter.
6. Describe the method of using an oscilloscope to display a waveform.

Syllabus

7.1 Types of instrument used for the measurement of AC, DC and RF voltages and currents; errors in measurement; analogue and digital multimeters, oscilloscopes.

7.2 Measurement of:
   (i) DC power input to power amplifiers;
   (ii) RF power output of power amplifiers;
   (iii) current at radio frequencies.

7.3 Purposes, operation and use of absorption wavemeters, crystal calibrators and digital frequency meters; accuracies.

7.4 Dummy loads, their construction and use in adjusting transmitters.

7.5 Use of standing-wave ratio meters.

7.6 Setting up and use of an oscilloscope to examine and measure waveforms and monitor the depth of modulation.

APPENDIX 4

Practice multiple-choice questions

This appendix contains two sample question papers in the multiple-choice format as defined in Chapter 1. These sample items illustrate the kind of question included in the examination. They are closely representative of the scope of the examination, but not exactly so as they were not originated by City and Guilds of London Institute.

The first paper contains 45 questions and the second 55 questions. The times are respectively 75 and 90 minutes. There is usually a break of 15 minutes between parts 1 and 2. A list of the answers will be found after the question papers.

Sample examination 1, Paper 1

Licensing conditions, transmitter interference and electromagnetic compatibility

1. If the licence is revoked, the Validation Document shall be:
   (a) destroyed
   (b) surrendered to the Secretary of State
   (c) returned to the local post office
   (d) returned to the RSGB

2. The Class B licence does not authorise the use of frequencies for transmitting:
   (a) above 144MHz
   (b) above 430MHz
   (c) in the microwave range
   (d) below 30MHz

3. The callsign GW4xxx is issued to:
   (a) a Class A licensee living in Wales
   (b) a Class B licensee living in Wales
   (c) a Class A licensee living in Winchester
   (d) a Class A licensee living in Scotland

4. The licence also authorises the licensee to operate in any country:
   (a) in the world
   (b) within the Commonwealth
   (c) which has implemented the appropriate CEPT recommendation
   (d) in Europe

5. Which of the following represents a valid log?
   (a) a loose-leaf book
   (b) a non-loose-leaf book
   (c) a magnetic disc containing propagation and RTTY programs
   (d) a magnetic tape which also includes games programs

6. Providing the licence fee is paid, an amateur licence:
   (a) is in force from year to year unless revoked
   (b) is valid for six months
   (c) is valid for five years
   (d) none of these

7. Operation from which of the following is not permissible under the licence conditions?
   (a) a private yacht on Loch Lomond
   (b) a private yacht between Liverpool and Northern Ireland
   (c) a helicopter above the Channel Isles
   (d) the inter-island ferry in the Orkney Islands

8. A log must be kept for:
   (a) mobile operation
   (b) pedestrian operation
   (c) main station address and all temporary locations
   (d) main station address only

9. Providing notice has been given to the Manager of the Radio Investigation Service office in whose area operation is to take place, the suffix to be used at the temporary location is:
   (a) /M
   (b) /MM
   (c) /P
   (d) not required

10. The transmission defined as RTTY shall encompass:
    (a) only ASCII transmissions
    (b) only binary-coded decimal encoding
    (c) AMTOR only
    (d) radio teletype and AMTOR

11. Using voice modulation, the nomenclature J3E corresponds to:
    (a) SSB with full carrier

(b) SSB with suppressed carrier
(c) FM using voice modulation
(d) a CW transmission

12. It is an offence to send by wireless telegraphy:
(a) certain misleading messages
(b) severe weather warnings
(c) test transmissions
(d) ASCII code

13. Which of the following bands are shared with other services?
(a) 3.5–3.8MHz
(b) 7.0–7.1MHz
(c) 14.0–14.35MHz
(d) 21.0–21.45MHz

14. If a station is located within 1km of the boundary of an airfield, the height of the antenna system above ground level must not exceed:
(a) 10 metres
(b) 15 metres
(c) 20 metres
(d) 50 metres

15. What is the peak envelope power allowed on the 432–440MHz band?
(a) 15dBW
(b) 26dBW
(c) 28dBW
(d) 40dBW

16. The licence states that the emitted frequency from the apparatus comprised in the station is as stable and free from unwanted emissions as:
(a) can be provided by resistor networks for amateur radio
(b) only suitable synthesisers can provide
(c) only suitable semiconductors for amateur radio permit
(d) the state of technical development in amateur radio apparatus reasonably permits

17. The first odd harmonic of 144.69MHz is:
(a) 48.23MHz
(b) 289.69MHz
(c) 434.07MHz
(d) 723.45MHz

18. If a transmitter is overdriven it is likely to cause:
(a) harmonics
(b) sub-harmonics
(c) a change in the modulation mode
(d) small DC variations

19. Which of the following might be effective at reducing risk of parasitic oscillations in a low-power VHF output stage?
(a) ferrite beads on the emitter lead of the power device
(b) ferrite beads on the microphone cable
(c) ferrite beads in series with the microphone
(d) ferrite beads on the loudspeaker leads

20. Interference is experienced on the 144MHz band from some 432MHz crystal-controlled equipment. The basic oscillator is around 12MHz. The most likely multiplication order is:
(a) $\times 2 \times 3 \times 3 \times 2$
(b) $\times 2 \times 2 \times 3 \times 3$
(c) $\times 3 \times 3 \times 2 \times 2$
(d) $\times 3 \times 2 \times 3 \times 2$

21. To stop unwanted radiations from an oscillator, it should be:
(a) enclosed in a metal box
(b) left unscreened
(c) not be RF decoupled
(d) placed in a paper box

22. Which of the plots below represents a filter suitable for following a microphone in order to limit the bandwidth?

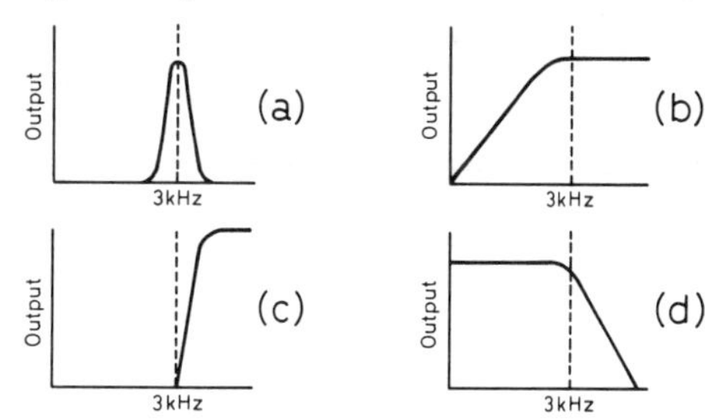

23. Which of the following waveforms is likely to give minimum interference due to key-clicks?

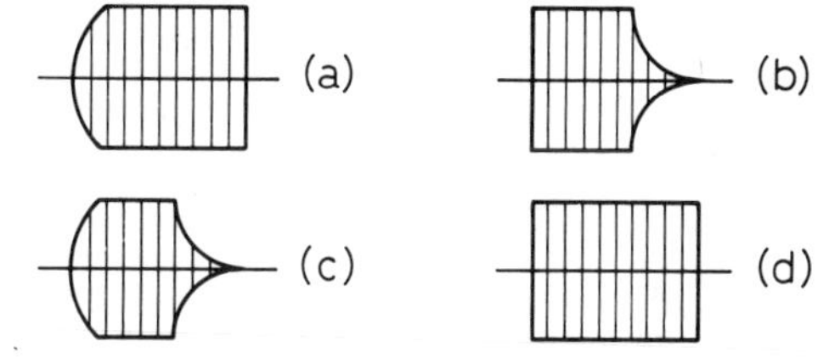

24. Chirp is a form of frequency instability. It is caused by:
(a) background noise
(b) overmodulation
(c) over-deviation
(d) pulling of an oscillator when keying

25. To minimise interference on adjacent channels, voice frequencies should be kept below:
(a) 500Hz
(b) 1kHz
(c) 3kHz
(d) 5kHz

26. If the frequency stability of a transmitter is poor it may cause:
(a) electric shocks
(b) operation out of band
(c) excessive collector dissipation
(d) excessive power to be drawn from the supply

27. If the accuracy of a digital readout on a transmitter is 10 parts per million, a frequency shown of 14.250MHz could be as high as:
    (a) 14.250001425MHz
    (b) 14.25001425MHz
    (c) 14.2501425MHz
    (d) 14.251425MHz

28. Which type of mixer keeps unwanted outputs to a minimum?
    (a) a balanced mixer
    (b) a product detector
    (c) single transistor mixers
    (d) single diode mixers

29. The bandwidth of data transmission should be kept to that of telephony, in order to:
    (a) help demodulation
    (b) conserve bandwidth
    (c) reduce transmitter power
    (d) reduce self-oscillation

30. An absorption wavemeter can be used to check for:
    (a) over-modulation
    (b) receiver overloading
    (c) band edge signals
    (d) correct selection of a harmonic from a multiplier circuit

31. The use of indoor transmitting antennas will:
    (a) be encouraged
    (b) never couple into the mains
    (c) have a good chance of coupling into the mains
    (d) give more long-distance contacts

32. A neighbour's hi-fi system is suffering from RF breakthrough. One possible cure would be:
    (a) ferrite beads on the transmitter lead
    (b) a capacitor across the transmitter lead
    (c) screened wire for the loudspeaker leads
    (d) open-wire feeder for the transmitter

33. A neighbour's TV is suffering RF breakthrough when you transmit on the 144MHz band. The TV set has a set-top antenna. There is no problem caused to your TV which receives the same station but uses a 10-element beam on the chimney. A possible cure is:
    (a) an external antenna for the neighbour's TV
    (b) a set-top antenna for your TV
    (c) a preamplifier between the neighbour's set-top antenna and their TV
    (d) use of twin feeder for the neighbour's set-top TV antenna

34. Which of the following is most likely to produce broadband continuous RF breakthrough:
    (a) an electric light switch
    (b) an incandescent bulb
    (c) a microwave transmitter
    (d) poor commutation in an electric drill

35. Before explaining to someone that the cause of RF breakthrough is lack of immunity in their equipment, one should:
    (a) make sure that there is no breakthrough on one's own domestic equipment
    (b) disconnect all one's own equipment from the mains
    (c) write a letter to the DTI
    (d) ignore all complaints

36. When operating a mobile HF set at home from a battery supply using the base antenna, there is no breakthrough problem. When using the same arrangement with an earthed battery charger also connected, breakthrough occurs on an electronic organ. The possible cause is:
    (a) the production of sub-harmonics at the transmitter
    (b) very strong received signals
    (c) poor RF earthing
    (d) RF earthing is too good

37. A transmitter has a power output of 100W. This is connected to an antenna of gain 11dB by a coaxial cable of loss 1dB. The ERP is:
    (a) 11W
    (b) 111W
    (c) 1000W
    (d) 2000W

38. It is found that interfering signals are being induced on the braid of an antenna downlead to a domestic FM radio by a 144MHz transmitter. One possible solution is:
    (a) to fit a braid-breaker filter on the antenna downlead
    (b) remove 144MHz transmitter earth connection
    (c) to increase 144MHz power output
    (d) fit 144MHz transmitter with a low-pass filter

39. A 432MHz amateur station causes breakthrough to a nearby TV receiver. Which of the following filters could be fitted in the TV downlead in order to minimise the breakthrough problem?

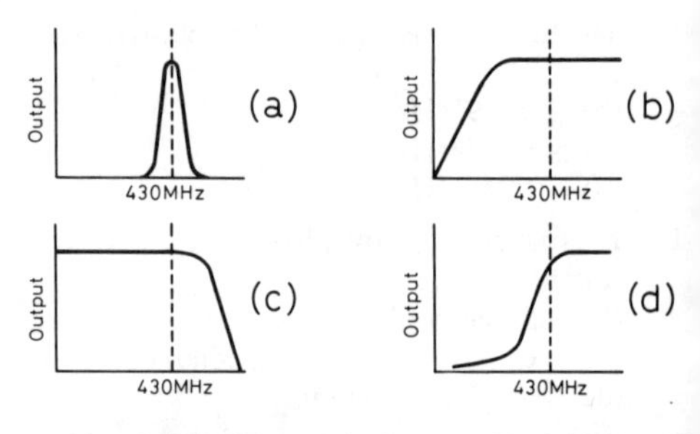

40. When making a filter to minimise breakthrough and cutting the coaxial cable to solder to the connecting socket, one should:
    (a) cut as much screening braid off as possible
    (b) cut as little screening braid off as possible
    (c) earth the centre conductor
    (d) solder screen and centre conductor together

41. In considering the equipment and power levels in a densely populated neighbourhood, it might be advisable to:
    (a) keep the antenna as low as possible
    (b) sit the antenna as remotely as possible from neighbours
    (c) use maximum ERP possible on 432MHz
    (d) always use long wire feeds
42. If a neighbour complains of breakthrough, one's immediate response should be:
    (a) blame the neighbour's equipment
    (b) tell them you will do nothing about it
    (c) stay polite and get them to help you investigate
    (d) to inform the DTI and RSGB
43. In making decisions on type of feeder for a transmitting antenna, then to minimise the risk of interference:
    (a) consider only long wire feeds
    (b) use unscreened feeders near the building
    (c) use only screened feeders near a building
    (d) do not earth any part of the feeders
44. If the use of filters in a TV downlead fails to cure a breakthrough problem and the transmitter output is clean and the TV is still under guarantee, then:
    (a) suggest the complainant returns the TV to the supplier and tell them it lacks immunity and ask for it to be put right
    (b) try and cure the problem internally in the TV
    (c) try filters in the transmitter output feed
    (d) tell the complainant not to use the TV
45. An amateur radio transmitter/antenna system has an ERP of 100W; the field strength at a distance of 100m in free space is about:
    (a) 0.35V/m
    (b) 0.7V/m
    (c) 3.5V/m
    (d) 7.02V/m

Sample examination 1, paper 2

Operating practices, procedures and theory

1. Having established contact on a calling frequency it is good practice to:
    (a) stay on the same frequency
    (b) move to another frequency
    (c) invite others to join in on the same frequency
    (d) be objectionable to all others calling
2. The purpose of a terrestrial repeater is to
    (a) increase satellite coverage
    (b) increase the range of mobile stations
    (c) increase the range of fixed stations
    (d) minimise contacts by pedestrian stations
3. The Q-code for 'standby' is:
    (a) QRN
    (b) QRM
    (c) QRS
    (d) QRX
4. COIL using the phonetic alphabet would be:
    (a) Charlie, Oscar, India, Lima
    (b) Charlie, Ocean, Italy, Lima
    (c) Coil, Oscar, Inductance, London
    (d) Charlie, Oscar, Italy, London
5. The Band Plans should be observed because:
    (a) they are mandatory
    (b) they are governed by international regulations
    (c) they are intended to aid operating
    (d) they are only for novices
6. It is good safety practice to:
    (a) use plastic piping for earthing
    (b) unearth all metal cases
    (c) have no master switch
    (d) supply all mains power via a master switch
7. When calling a station it is good practice to:
    (a) put your callsign first
    (b) use your callsign only
    (c) put the callsign of the station being called first
    (d) use the callsign of the other station only
8. In the RST code T is for:
    (a) temperature of PA stage
    (b) tone
    (c) time of transmission
    (d) transmitter type
9. To prevent annoying other users on a band, a transmitter should always be tuned initially:
    (a) on a harmonic outside the band
    (b) into an antenna
    (c) into a dummy load
    (d) into a dipole
10. Two 10kΩ resistors are connected in parallel across a 5V DC supply. The total current taken is:
    (a) 50μA
    (b) 0.5mA
    (c) 1mA
    (d) 1A
11.

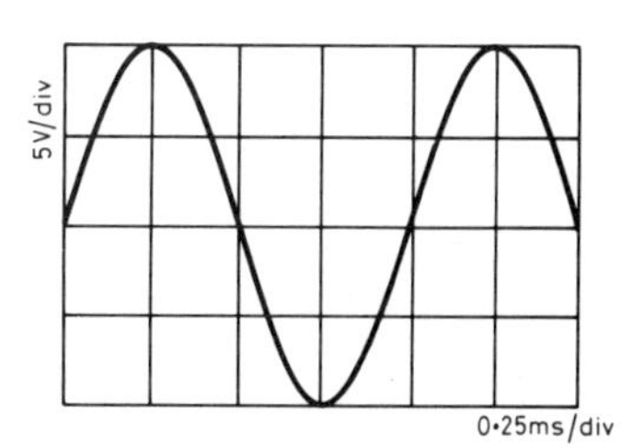

The above diagram represents a trace on an oscilloscope. What is the frequency of the displayed waveform?
    (a) 1kHz
    (b) 5kHz

(c) 10kHz
(d) 100kHz

12. In the diagram in question 11, what is the peak to peak value of the waveform?
(a) 1V
(b) 2V
(c) 10V
(d) 20V

13. As the frequency rises, the reactance of an inductor:
(a) stays constant
(b) decreases
(c) increases
(d) does none of these

14.

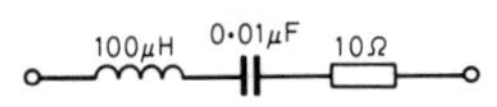

The resonant frequency of the above circuit is:
(a) 1.59155kHz
(b) 15.9155kHz
(c) 159.155kHz
(d) 1591.55kHz

15. A power gain of 4 is equivalent to:
(a) 3dB
(b) 6dB
(c) 10dB
(d) 16dB

16. In the following diagram, which represents the diode in a conducting condition?

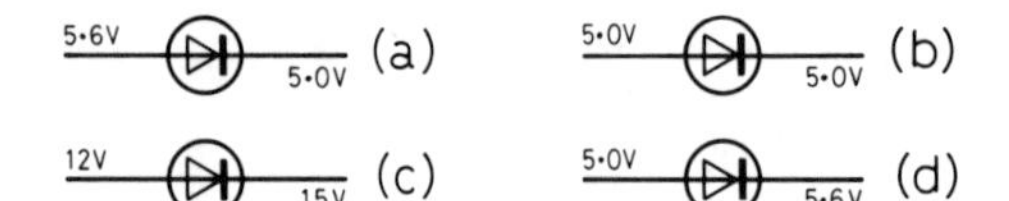

17. If the base potential of an NPN transistor is held at the emitter potential, the collector current will be:
(a) zero
(b) always 1A
(c) between 10mA and 2A
(d) very high

18. The output impedance of an emitter-follower buffer amplifier is:
(a) infinite
(b) very high
(c) 0
(d) fairly low

19. The circuit shown below depicts:
(a) an audio amplifier
(b) an RF amplifier
(c) a mixer
(d) a BFO

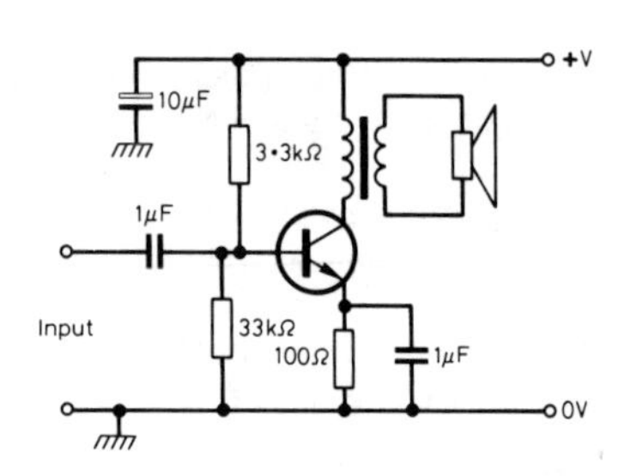

20. The circuit in question 19 operates in:
(a) Class A
(b) Class AB
(c) Class B
(d) Class C

21. Integrated circuits that performs logic functions with binary signals are classified as:
(a) linear circuits
(b) linear amplifiers
(c) analogue mixers
(d) digital circuits

22.

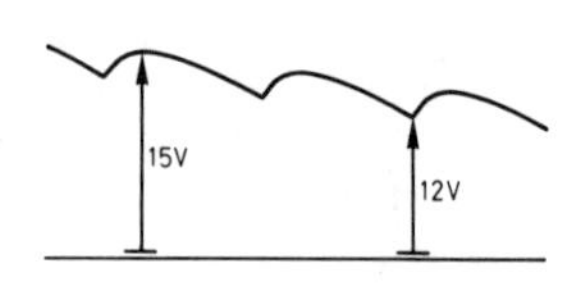

The trace shown above is from a power supply. The ripple is:
(a) 1V
(b) 3V
(c) 12V
(d) 15V

23. The frequency difference between the wanted RF signal and the so-called second channel is:
(a) twice the third IF
(b) twice the wanted RF
(c) twice the first IF
(d) the wanted RF plus the first IF

24. In a direct-conversion receiver, the local oscillator is:
(a) very much higher than the received signal
(b) very much lower than the received signal
(c) always at 10.7MHz
(d) very close to that of the received frequency

25. The squelch in a receiver is operated by:
(a) the VFO
(b) the power supply
(c) the heterodyne oscillator
(d) either the IF or AF signal

26. The effect of the AGC on receipt of a very strong incoming signal is to:
(a) reduce the VFO output
(b) reduce the gain of the RF and IF amplifiers
(c) reduce the power supply voltage
(d) reduce a filter response

27. A high first IF allows easier filtering to prevent:
    (a) power supply ripple
    (b) local oscillator breakthrough
    (c) second-channel interference
    (d) second IF breakthrough

28.

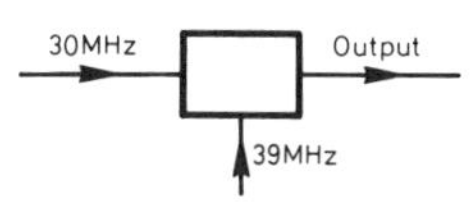

The principal outputs from the above circuit are:
    (a) 9 and 39MHz
    (b) 9 and 69MHz
    (c) 30 and 39MHz
    (d) 39 and 69MHz

29. In order to demodulate CW transmissions in an AM-only receiver, which of the following is required?
    (a) a BFO
    (b) an FM detector
    (c) a crystal multiplier
    (d) a morse key

30. The coil determining the frequency of a transmitter VFO should be:
    (a) air cored with no former
    (b) air cored and tightly wound on a former
    (c) air cored and loosely wound on a former
    (d) wound on a metallic former

31. To produce double sideband, suppressed carrier signal, which of the following should be used?
    (a) a balanced mixer
    (b) a crystal filter
    (c) a single transistor mixer
    (d) a single diode mixer

32.

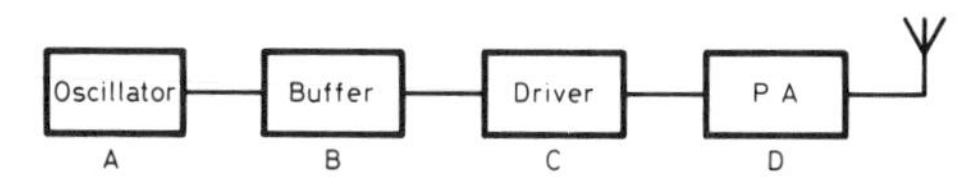

The above represents a CW transmitter. Ideally, keying should be applied at:
    (a) A
    (b) B
    (c) C
    (d) D

33. The output power from an FM transmitter:
    (a) is constant irrespective of modulation
    (b) varies with modulation
    (c) is zero with no modulation
    (d) reduces to 50% with modulation

34. The output amplifier of an SSB transmitter must:
    (a) act as a switch
    (b) be in a linear mode
    (c) be in a non-linear mode
    (d) act as a multiplier

35.

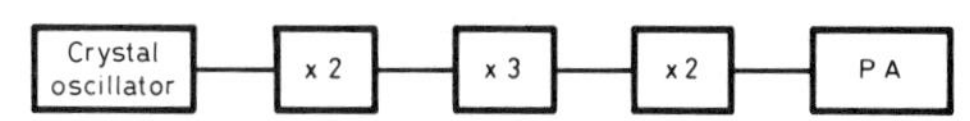

In the circuit arrangement above, to give an output of 145.000MHz, the crystal oscillator is at:
    (a) 10.7142MHz
    (b) 12.08333MHz
    (c) 72.5000MHz
    (d) 145.000MHz

36. If a transmitter output impedance is 50Ω, for optimum power transfer the load should be:
    (a) 50Ω
    (b) 75Ω
    (c) 100Ω
    (d) 150Ω

37. The RF spectrum required to transmit an SSB signal is:
    (a) half the modulating signal bandwidth
    (b) the same as the modulating signal bandwidth
    (c) twice the modulating signal bandwidth
    (d) zero

38. The wavelength of a signal of 100MHz in free space is:
    (a) 30mm
    (b) 0.3m
    (c) 3.0m
    (d) 30m

39. In order to radiate, an electromagnetic wave must have:
    (a) E field only
    (b) H field only
    (c) E and H field
    (d) air to travel in

40. Polarisation of an electromagnetic wave is fixed by:
    (a) the direction of the H field
    (b) the direction of propagation
    (c) by an anti-phase signal
    (d) the orientation of the transmitting antenna

41. The signals returned from the layers above the earth are referred to as:
    (a) the ground wave
    (b) the ionospheric wave
    (c) the tropospheric wave
    (d) the direct wave

42. As frequency increases, the ionisation to reflect a signal back to the earth must:
    (a) decrease
    (b) go to zero
    (c) not change
    (d) increase

43. Fading can be caused by:
    (a) a poor antenna
    (b) horizontal polarisation
    (c) interaction of the sky and ground wave
    (d) poor coaxial cable

44. A transmission line is said to be perfectly matched when it is terminated by a resistance equal to:
    (a) half the characteristic impedance
    (b) twice the characteristic impedance
    (c) an open-circuit
    (d) the characteristic impedance

45.

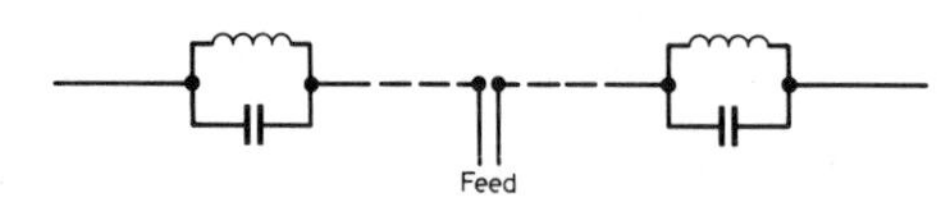

The above antenna represents:
    (a) an end-fed wire
    (b) a simple dipole
    (c) a cubical quad
    (d) a trap dipole

46. The addition of reflectors and a director to a folded dipole:
    (a) raises its impedance
    (b) has no effect on its impedance
    (c) lowers its impedance
    (d) produces no directivity

47. The typical accuracy of a moving coil meter is:
    (a) 0.03%
    (b) 0.3%
    (c) 3%
    (d) 10%

48. A dummy load for use at VHF should be made from:
    (a) wire-wound resistors
    (b) carbon resistors
    (c) metal oxide resistors
    (d) electric-fire heating elements

49.

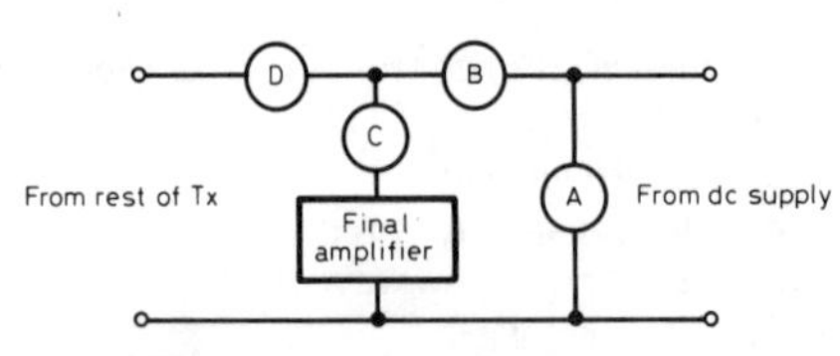

To measure DC input power to a final amplifier, various meters are used. Those necessary are:
    (a) voltmeter at A, ammeter at B
    (b) voltmeter at A, ammeter at C
    (c) voltmeter at C, ammeter at A
    (d) voltmeter at C, ammeter at B

50. The action of the so-called 'dip oscillator' depends on:
    (a) extraction of energy from the tuned circuit on test
    (b) extraction of energy from the dip oscillator by circuit under test
    (c) radiation from a nearby transmitter
    (d) the tuned circuit on test changing the dip oscillator frequency

51.

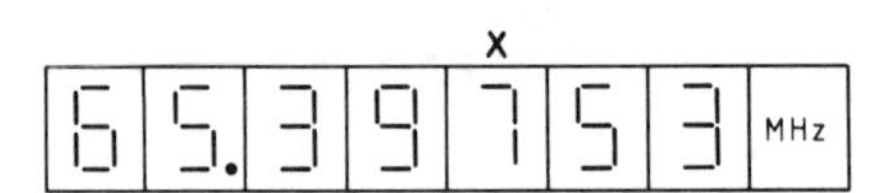

The above represents the output of a digital frequency counter. What does the digit at X signify?
    (a) hertz
    (b) tens of hertz
    (c) hundreds of hertz
    (d) thousands of hertz

52. The coaxial cable from an SWR meter to an antenna develops a fault so that so that no power reaches the antenna the SWR meter will read:
    (a) zero
    (b) 1:1
    (c) high
    (d) very low

53. Which of the following would be used to examine the shape of a waveform?
    (a) an oscilloscope
    (b) an absorption wavemeter
    (c) a digital frequency counter
    (d) a dip meter

54. For accurate frequency measurement, which of the following instruments would be used?
    (a) an absorption wavemeter
    (b) an oscilloscope
    (c) a moving-coil multimeter
    (d) a digital frequency counter

55.

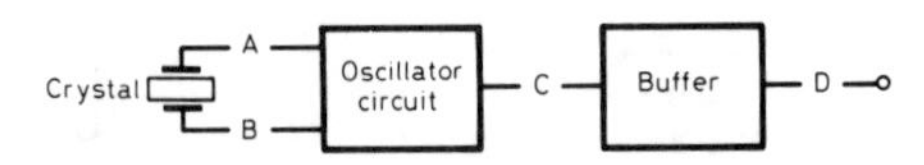

When measuring frequency in the above equipment, the probe should preferably be placed at:
    (a) A
    (b) B
    (c) C
    (d) D

Sample examination 2, paper 1

Licensing conditions, transmitter interference and electromagnetic compatibility

1. Which of the following is not defined as a user service?
    (a) fire brigade
    (b) British Telecom
    (c) County Emergency Planning Officer
    (d) St John Ambulance Service

2. An Amateur Licence A requires:
   (a) a pass in the RAE and the morse test
   (b) a pass in the RAE only
   (c) a pass in the morse test only
   (d) to have held a Class B licence previously

3. The nationality requirement for holding an amateur licence in the UK is:
   (a) British
   (b) British and Commonwealth
   (c) European
   (d) not specified

4. As well as amateur frequency transmissions, the licence allows reception of:
   (a) diplomatic messages
   (b) standard frequency transmissions
   (c) news agency transmissions
   (d) Police transmissions

5. Which of the following occurrences need not be entered into the station log?
   (a) tests for interference
   (b) station used by licensed operator other than licensee
   (c) station operated at temporary location
   (d) station temporarily dismantled

6. The callsign prefix GD should always be used whenever the station is being operated from:
   (a) Scotland
   (b) Isle of Man
   (c) Isle of Dogs
   (d) Guernsey

7. An English station while cycling in Wales should use the callsign:
   (a) G6xxx/P
   (b) G6xxx/GW/M
   (c) GW6xxx/M
   (d) GW6xxx/P

8. The 10 metre band lies between:
   (a) 28.0–28.7MHz
   (b) 28.7–29.0MHz
   (c) 28.0–29.0MHz
   (d) 28.0–29.7MHz

9. The licence requires that the purity of the transmitter output should be checked:
   (a) weekly
   (b) from time to time
   (c) daily
   (d) never

10. In the UK, a licensee may receive messages from an overseas amateur on a frequency band not specified in the first column of the schedule as long as the licensee:
    (a) transmits only in a band specified in the schedule
    (b) obtains a permit from the Licensing Authority
    (c) holds a reciprocal licence from that country
    (d) informs the RSGB

11. Time in the log book must always be in:
    (a) local time
    (b) UTC
    (c) BST
    (d) UTC + 1 hour

12. When the station is used from a temporary location and the Radio Investigation Service not informed, the location must be:
    (a) given at least every five minutes
    (b) sent in CW only
    (c) given within 0.5km
    (d) given within 5km

13. When using phonetics the licence document specifies that the phonetic alphabet:
    (a) should be used
    (b) is not necessary
    (c) is mandatory
    (d) must not be used

14. The amateur licence can be revoked by a broadcast by:
    (a) the BBC
    (b) the IBA
    (c) the RSGB
    (d) the amateur

15. Codes and abbreviations may be used by the licensee as long as:
    (a) they are specified only by the RSGB
    (b) they are in secret cypher
    (c) they do not obscure the meaning of the communication
    (d) Q-codes only are used

16. Over-modulation of an AM signal is likely to cause:
    (a) excessive deviation
    (b) 10 sidebands
    (c) minimum interference
    (d) severe splatter on adjacent frequencies

17. Power supplies to RF power amplifiers should:
    (a) be open wires
    (b) be AF filtered
    (c) be RF filtered
    (d) be inductively coupled

18. If an L-C oscillator is used to generate directly a signal at 14.05MHz for a CW transmitter and it drifts by −1%, it will:
    (a) stay within the designated band
    (b) go above the top band edge
    (c) go below the bottom band edge
    (d) be rejected

19. To check the calibration of a transceiver with a VFO at the band edges, the minimum equipment needed is:
    (a) a dip meter
    (b) a crystal-controlled digital frequency counter
    (c) an absorption wavemeter
    (d) an oscilloscope

20. Parasitic oscillations can cause interference. They are:
    (a) of very low frequency
    (b) always twice the operating frequency
    (c) high in frequency but not related to the operating frequency
    (d) always three times the operating frequency
21. Any non-linear device will produce:
    (a) mixing products
    (b) amplification
    (c) filtering
    (d) key-clicks
22. To minimise unwanted radiation of sub-harmonics and harmonics, a VHF transmitter should be followed by:
    (a) a low-pass filter
    (b) a bandpass filter
    (c) a high-pass filter
    (d) a notch filter
23. A domestic receiver, having an IF of 455kHz and receiving a signal on 945kHz, experiences strong breakthrough from someone on the 160m band. This could be caused by second-channel interference of:
    (a) 1.810MHz
    (b) 1.825MHz
    (c) 1.835MHz
    (d) 1.855MHz
24. Which of the following filter characteristics represents a filter for following an all-band HF transceiver?

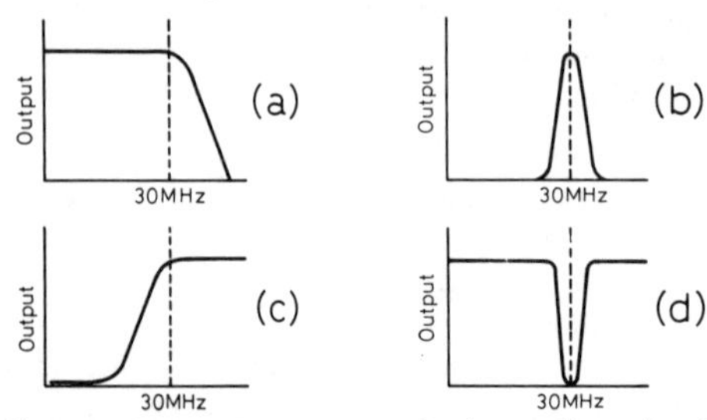

25. Spurious resonances may occur in decoupling circuits. This is due to:
    (a) the power supply
    (b) self-resonance of RF chokes
    (c) saturation of the core of RF chokes
    (d) the resistive element of an RF choke
26. If the coil in a VFO has no former, then vibrations:
    (a) will keep the frequency generated in the band
    (b) may take the frequency generated out of band
    (c) will provide a clean signal
    (d) are beneficial
27. So as not to cause unnecessary sideband splatter the percentage modulation of an AM signal must be kept below:
    (a) 25%
    (b) 50%
    (c) 75%
    (d) 100%
28. The minimum equipment to check for harmonic radiation from a transmitter is:
    (a) a digital frequency counter
    (b) an absorption wavemeter
    (c) a multimeter
    (d) an oscilloscope
29. An ammeter in the PA stage of a transmitter shows slight fluctuations when the transmitter is not being keyed. This possibly indicates:
    (a) the presence of parasitic oscillations
    (b) good biasing arrangement
    (c) the reception of an interfering signal
    (d) electromagnetism
30. The station shall be constructed and maintained so as:
    (a) not to cause any undue interference
    (b) to cause interference with wireless telegraphy
    (c) to cause interference to televisions only
    (d) to produce spurious and harmonic radiations
31. The type of interference caused by a transmitter can be classified as:
    (a) broad-band interference
    (b) narrow-band interference
    (c) pseudo-random interference
    (d) white noise
32. The fifth harmonic of a 145MHz transmission lies in:
    (a) a Police band
    (b) an FM radio broadcast band
    (c) a UHF TV band
    (d) the microwave region
33. A 435MHz high-gain transmitting antenna points right at a UHF TV receiving antenna. This could cause:
    (a) problems with the 435MHz receiver
    (b) overloading of the TV front-end
    (c) self-oscillation of the 435MHz transmitter
    (d) melting of the TV antenna elements
34. Pick-up in the IF stage of a TV set usually results in:
    (a) problems with the picture
    (b) poor power supply regulation
    (c) random channel changing
    (d) no audio output
35. If the mains earth is used as an RF earth, this could be prone to causing:
    (a) mains hum
    (b) mains-borne interference
    (c) parasitic oscillations
    (d) self-oscillation
36. Capacitors used for bypassing at RF should be:
    (a) electrolytic
    (b) paper
    (c) polycarbonate
    (d) ceramic

37. Which of the following would be useful in rejecting an unwanted signal at the input of a receiver?

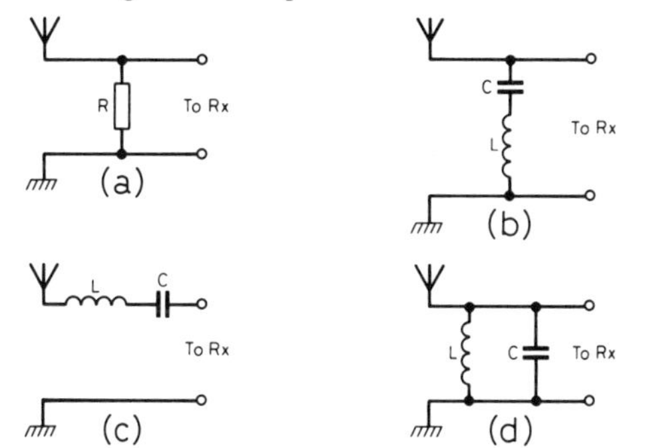

38. A braid-breaking choke in a TV antenna downlead will block:
    (a) all AC signals
    (b) out-of-phase interfering signals
    (c) in-phase interfering signals
    (d) mains hum

39. A corroded connector on a neighbour's TV receiving antenna may cause:
    (a) unwanted mixing products due to it exhibiting diode properties
    (b) mains rectification
    (c) enhanced signal reception due to its filtering properties
    (d) increased amplification

40. A radio transmitting system has an ERP of 225W. At what distance away is the field strength 0.5V/m?
    (a) 0.21m
    (b) 2.1m
    (c) 21m
    (d) 210m

41. To minimise RF going back into the mains, which of the following could be fitted in the mains input of a piece of equipment?

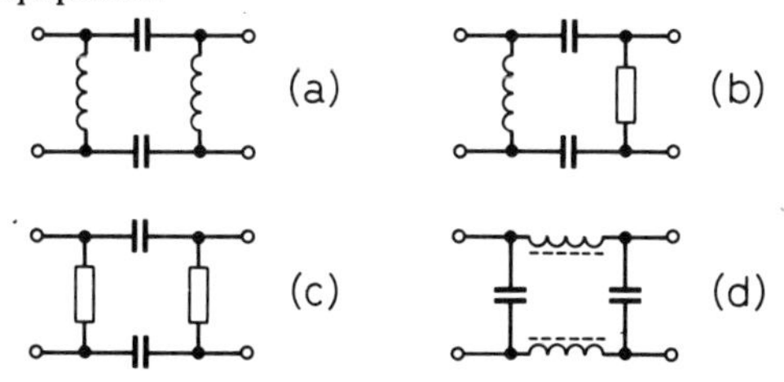

42. If an antenna runs close and parallel to an overhead 240V power line, there may be the possibility of:
    (a) harmonic generation
    (b) 50Hz modulation on all signals
    (c) producing mains-borne interference
    (d) cheap power

43. A transmitter is connected by a very short coaxial cable to a 6dB collinear antenna. When the output power to the antenna is reduced to 5W no more interference is caused to a neighbour's hi-fi system. This corresponds to an effective radiated power of:
    (a) 1W
    (b) 10W
    (c) 11W
    (d) 20W

44. The best place for an HF beam to minimise interference for an amateur living in a semi-detached house is:
    (a) on the joint chimney stack in the centre of the house
    (b) overhanging next door's roof space
    (c) as high and far away as possible
    (d) as low and far away as possible

45. A neighbour complains of breakthrough to their television but says it goes when they disconnect the antenna. This also coincides with your transmission times. As a first step:
    (a) try a mains filter
    (b) suggest they use a set-top antenna
    (c) try a filter in the TV downlead
    (d) renew the antenna cable

Sample examination 2, Paper 2

Operating practices, procedures and theory

1. Before making a CQ call:
    (a) listen on the frequency before commencing
    (b) send a series of Vs in morse
    (c) send a 1750Hz tone
    (d) keep giving your callsign

2. To access a repeater in the UK one must:
    (a) send the callsign of the repeater in ASCII
    (b) send a 1750Hz tone burst
    (c) send a 1850Hz tone burst
    (d) speak the callsign of the repeater

3. If a station asks 'please QSY', this means:
    (a) there is fading
    (b) change frequency
    (c) stop transmitting
    (d) reply in morse

4. The only general call allowed from an amateur station is:
    (a) a news bulletin
    (b) a CQ call
    (c) a third-party call
    (d) on VHF

5. When working through a satellite use:
    (a) as much power as possible
    (b) Esperanto
    (c) sufficient power to maintain reliable communications
    (d) FM only

6. For safety reasons all exposed metal work in an amateur station should be:
   (a) connected to mains neutral
   (b) free of earth connections
   (c) left completely floating
   (d) connected to a good RF earth
7. If a readability report of 2 is given, this would indicate:
   (a) unreadable
   (b) only readable with considerable difficulty
   (c) readable with practically no difficulty
   (d) perfectly readable
8. When wearing headphones it is not advisable to:
   (a) be calling CQ
   (b) have one's hands inside live equipment
   (c) be switching off
   (d) have rubber gloves on
9. Which of the following uses the International Phonetic Alphabet?
   (a) Boston, Uruguay, Gordon
   (b) Belgium, Units, Gravity
   (c) Bee, You, Gee
   (d) Bravo, Uniform, Golf
10. A light bulb is rated at 12V, 3W. The current drawn when used on a 12V source is:
   (a) 250mA
   (b) 750mA
   (c) 4A
   (d) 36A
11. The tolerance of a resistor is given as 10%. If the nominal value is 4700Ω, then its value must lie between:
   (a) 4230 and 5170Ω
   (b) 4653 and 4747Ω
   (c) 4230 and 4747Ω
   (d) 4653 and 5170Ω
12.

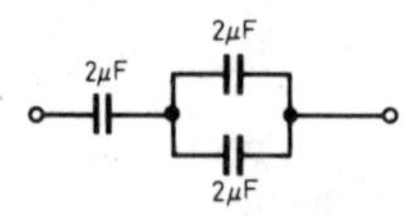

The total capacitance in the above circuit is:
   (a) 1.33μF
   (b) 3μF
   (c) 3.5μF
   (d) 6μF
13.

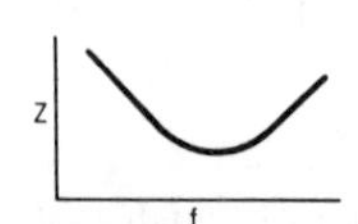

The above impedance-frequency curve represents:
   (a) a capacitance
   (b) a parallel tuned circuit
   (c) an inductance
   (d) a series-tuned circuit
14. The resonant frequency of a tuned circuit is 1MHz and the bandwidth at the −3dB points is 10kHz. The Q factor of the circuit is:
   (a) 50
   (b) 100
   (c) 500
   (d) 1000
15.

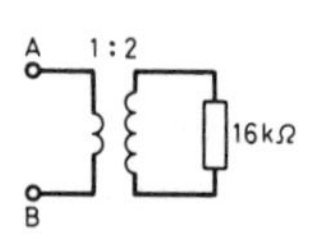

The impedance seen at terminals AB in the diagram above is:
   (a) 0Ω
   (b) 250Ω
   (c) 1000Ω
   (d) 4000Ω
16. There are two basic forms of transistor, these are:
   (a) PNP and NNP
   (b) PNP and NPN
   (c) PPN and NNP
   (d) NPP and PNN
17. A varactor diode acts like:
   (a) a variable resistance
   (b) a variable voltage regulator
   (c) a variable capacitance
   (d) a variable inductance
18.

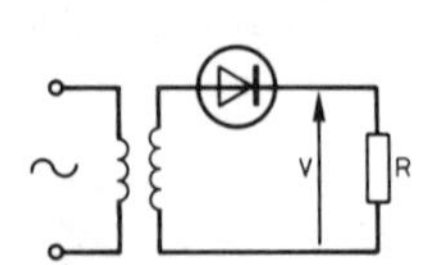

The above circuit is for:
   (a) half-wave rectification
   (b) full-wave voltage stabilisation
   (c) reverse bias protection
   (d) voltage multiplication
19. Colpitts, Hartley, Clapp-Gouriet and Vackar are all types of:
   (a) power supply
   (b) amplifier
   (c) oscillator
   (d) modulator
20. For a transistor to conduct:
   (a) the base must be 0.6V above the emitter
   (b) the base must be soldered to the emitter
   (c) the collector must be soldered to the emitter
   (d) the base lead must be removed
21. A Class C amplifier conducts over:
   (a) the complete cycle
   (b) three-quarters of a cycle

(c) exactly half a cycle
(d) less than half a cycle

22. The input resistance of a common-emitter amplifier stage is in the region of:
(a) 50Ω
(b) 250Ω
(c) 2kΩ
(d) 20kΩ

23. The main filtering of the wanted signal in a superheterodyne receiver is normally accomplished in:
(a) the RF amplifier
(b) the audio circuitry
(c) the second mixer
(d) the IF section

24. A receiver is quoted as having a sensitivity of 0.5μV for 20dB S/N ratio. The latter refers to a comparison between:
(a) the unwanted signal and wanted noise
(b) the wanted signal and unwanted noise
(c) the unwanted signal and unwanted noise
(d) the wanted signal and wanted noise

25.

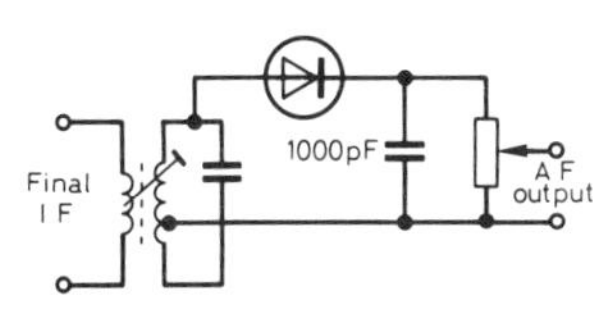

The circuit shown above is that of:
(a) an envelope detector
(b) a single-phase mains rectifier
(c) a discriminator
(d) a phase detector

26. In the detection of SSB signals there is normally a carrier insertion oscillator. In a high-quality receiver this is:
(a) a VFO
(b) varactor controlled
(c) inductor controlled
(d) crystal controlled

27. One advantage of FM over AM is:
(a) increased distance range
(b) narrower bandwidth required
(c) freedom from most sources of 'man-made' interference
(d) no antenna is needed

28. For CW reception, the difference in frequency of the BFO and final IF should be about:
(a) 1kHz
(b) 10kHz
(c) 455kHz
(d) 10.7MHz

29. In frequency modulation, the amplitude of the modulating signal manifests itself as:
(a) amplitude variations of the carrier
(b) envelope variations of the carrier
(c) deviation from average carrier frequency
(d) a constant carrier offset

30. The frequency of a VFO depends on an L-C circuit. The frequency is determined by:
(a) the difference between L and C
(b) the sum of L and C
(c) the ratio L/C
(d) the product of L and C

31.

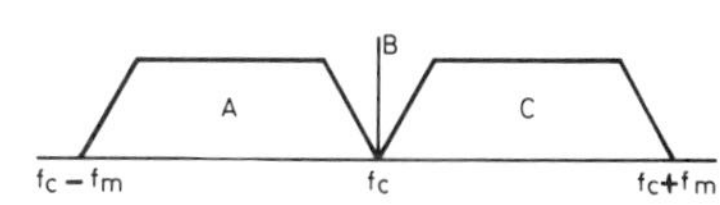

The above spectrum plot is typical of:
(a) amplitude modulation
(b) amplitude modulation, suppressed carrier
(c) single sideband, reduced carrier
(d) single sideband, full carrier

32.

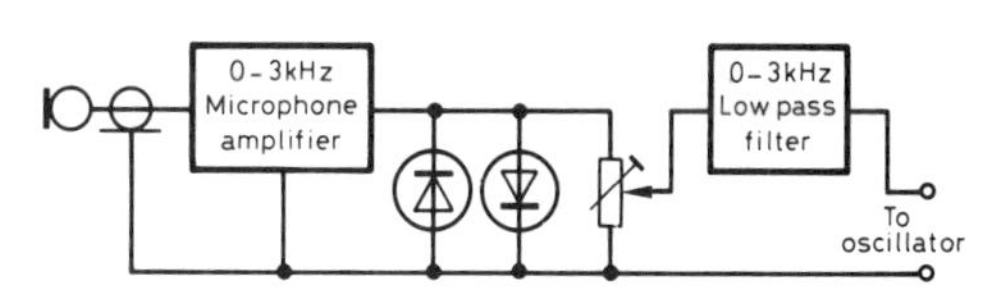

The above is part of a transmitter. It is typical of:
(a) an RF filter
(b) a deviation control
(c) an amplitude modulator
(d) an SSB generator

33. A VFO should ideally be followed by:
(a) a buffer amplifier
(b) a power amplifier
(c) a Class C amplifier
(d) a notch filter

34. Which of the following represents 100% amplitude modulation?

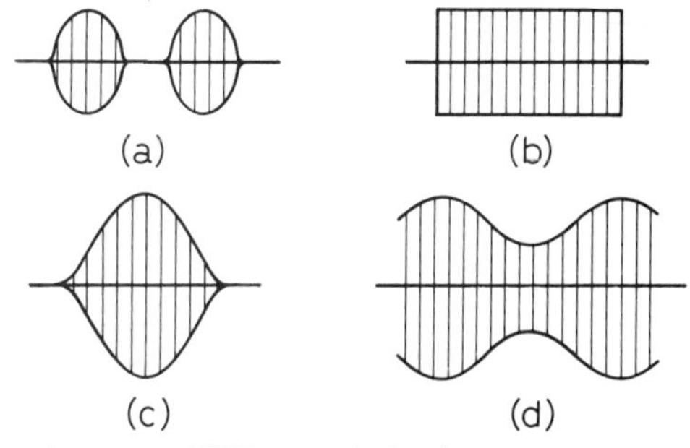

35. The advantage of J3E transmission is:
(a) power output minimised with modulation
(b) power output maximised with no modulation
(c) minimum power dissipation in PA with no modulation
(d) maximum power dissipation in PA with no modulation

36.

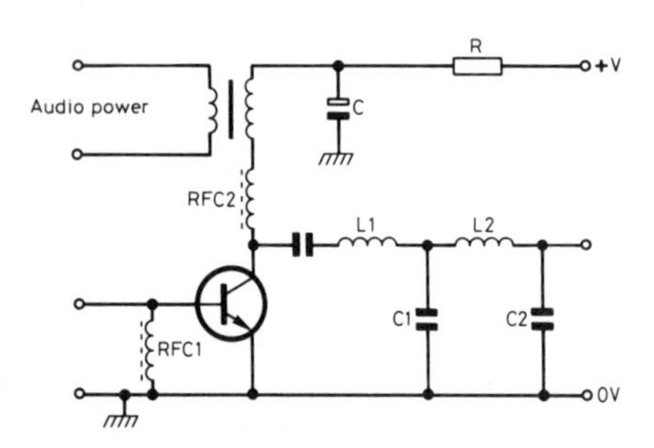

The above circuit represents a:
(a) frequency modulator
(b) amplitude modulator
(c) phase modulator
(d) amplitude demodulator

37. In the diagram of question 36, components L1, L2, C1 and C2 provide:
(a) harmonic suppression and impedance matching
(b) harmonic suppression only
(c) output matching only
(d) audio filtering

38. In electromagnetic radiation, which of the following is true?
(a) E and H are at 180° to each other
(b) E, H and the direction of propagation are all at right-angles to each other
(c) the angle between E and H is 0°
(d) the velocity of propagation is at 180° to the E field but in line with the H field

39. The major mode of propagation up to about 2MHz is by:
(a) direct wave
(b) tropospheric wave
(c) ionospheric wave
(d) ground wave

40. The skip zone is where the ground wave:
(a) is enhanced
(b) is reflected
(c) has diminished and the reflected wave has not returned to earth
(d) ground and reflected wave combine

41. Refraction of an electromagnetic wave is:
(a) the same as reflection
(b) the bending of its path
(c) absorption by the ionosphere
(d) bouncing from a stellar object

42. The MUF for a given radio path is:
(a) the mean of the maximum and minimum usable frequencies
(b) the maximum usable frequency
(c) the minimum usable frequency
(d) the mandatory usable frequency

43. A 10cm wavelength in air corresponds to:
(a) 3MHz
(b) 300MHz
(c) 3GHz
(d) 30GHz

44. If two signals arrive at a point out of phase:
(a) fading will occur
(b) signal enhancement occurs
(c) cross-polarisation is produced
(d) the antenna impedance varies

45. Inserting traps into each leg of a dipole:
(a) allows it to only operate on one band
(b) eliminates harmonics
(c) gives broad-band matching
(d) allows it to resonate on at least two bands

46. What is the length of a piece of coaxial cable cut for a full wavelength at 100MHz, if the velocity factor is 0.66?
(a) 0.198m
(b) 1.98m
(c) 3m
(d) 19m

47. A moving-coil meter depends on which of the following in order to operate?
(a) interaction of an electric and magnetic field
(b) interaction of a permanent and electromagnetic field
(c) the interaction of two permanent magnetic fields
(d) an electric field only

48. Which of the following items could be attached to a moving-coil meter in an attempt to measure power?
(a) a resistor
(b) a thermistor
(c) a thermocouple
(d) a therm

49. An SWR meter is inserted into a perfectly matched transmitter/antenna system. The value shown should indicate:
(a) 10W reflected power
(b) 1:1 SWR
(c) 1:0 SWR
(d) 0:1 SWR

50. Which of the instruments below has the highest accuracy?
(a) a heterodyne wavemeter
(b) a digital frequency counter
(c) an absorption wavemeter
(d) an oscilloscope

51. To check that a crystal is working on its correct overtone, the simplest piece of equipment necessary is:
(a) a voltmeter
(b) an ammeter
(c) an absorption wavemeter
(d) a dip oscillator

52.

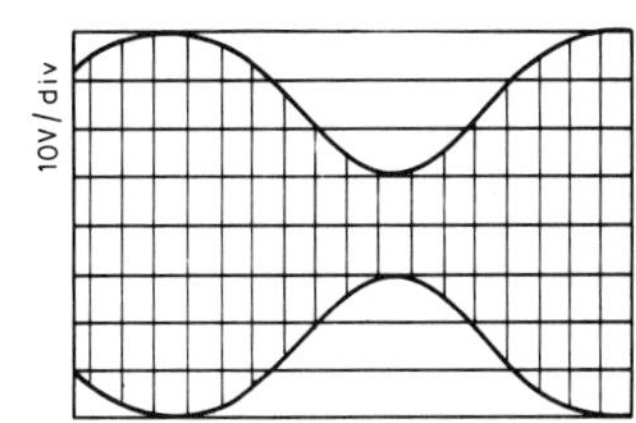

The above trace is on an oscilloscope monitoring an AM transmission. The depth of modulation is:
(a) 40%
(b) 50%
(c) 60%
(d) 70%

53. A good dummy load is constructed from:
(a) light bulbs
(b) a column of water
(c) wire-wound resistors
(d) non-reactive resistors

54. The typical accuracy of a dip oscillator might be:
(a) 0.001%
(b) 0.05%
(c) 1.0%
(d) 10%

55. An RTTY signal requires a bandwidth of ±3kHz. A frequency counter of accuracy 1 part per million is used to check the frequency readout of the 145MHz transmitter. How close can the signal be transmitted to the band edge to ensure transmission is within the licence conditions?
(a) 1.55kHz
(b) 2.855kHz
(c) 3.145kHz
(d) 4.45kHz

Answers

Question number	sample exam 1 paper 1	sample exam 1 paper 2	sample exam 2 paper 1	sample exam 2 paper2
1	b	b	b	a
2	d	b	a	b
3	a	d	d	b
4	c	a	b	b
5	b	c	d	c
6	a	d	b	d
7	c	c	c	b
8	c	b	d	b
9	d	c	b	d
10	d	c	a	a
11	b	a	b	a
12	a	d	d	a
13	a	c	a	d
14	b	c	a	b
15	b	b	c	d
16	d	a	d	b
17	c	a	c	c
18	a	d	c	a
19	a	a	b	c
20	b	a	c	a
21	a	d	a	d
22	d	b	b	c
23	c	c	d	d
24	d	d	a	b
25	c	d	b	a
26	b	b	b	d
27	c	c	d	c
28	a	b	b	a
29	b	a	a	c
30	d	b	a	d
31	c	a	b	a
32	c	b	c	b
33	a	a	b	a
34	d	b	a	c
35	a	b	b	c
36	c	a	d	b
37	c	b	b	a
38	a	c	c	b
39	d	c	a	d
40	b	d	d	c
41	b	b	d	b
42	c	d	c	b
43	c	c	d	c
44	a	d	c	a
45	b	d	c	d
46		c		b
47		c		b
48		b		c
49		b		b
50		b		b
51		d		c
52		c		c
53		a		d
54		d		d
55		d		c

Index

Get more out of amateur radio . . . as an RSGB member!

Radio Communication

An outstanding magazine, sent free of charge to all members, which covers a wide range of interests and which features the best and latest amateur radio news. There's technical articles, equipment reviews, and the famous 'Technical Topics' column to keep you up to date with technical matters. There are also regular news columns for HF, VHF/UHF, microwaves, SWL, satellites, data and contests. And, if you're after equipment, the Members' Ads offer the best bargains around.

QSL Bureau

Members enjoy the use of the QSL Bureau free of charge for both outgoing and incoming cards. This can save you a good deal of postage.

Specialised Equipment Insurance

Insurance for your valuable equipment which has been arranged specially for members. The rates are very advantageous.

Government Liaison

One of the most vital features of the work of the RSGB is the ongoing liaison with the UK Licensing Authority – presently the Radiocommunications Agency of the Department of Trade and Industry. Setting and maintaining the proper framework in which amateur radio can thrive and develop is essential to the well-being of amateur radio. The Society spares no effort in defence of amateur radio's most precious assets – the amateur bands.

Operating Awards

A wide range of operating awards are available via the responsible officers: their names can be found in the front pages of *Radio Communication* and in the Society's *Amateur Radio Call Book*. The RSGB also publishes a book which gives details of most major awards.

Contests (HF/VHF/Microwave)

The Society has two contest committees which carry out all work associated with the running of contests. The HF Contests Committee deals with contests below 30MHz, whilst events on frequencies above 30MHz are dealt with by the VHF Contests Committee.

Morse Testing

In April 1986 the Society took over responsibility for morse testing of radio amateurs in the UK. If you wish to take a morse test, write direct to RSGB HQ (Morse tests) for an application form.

Slow Morse

Many volunteers all over the country give up their time to send slow morse over the air to those who are preparing for the 5 and 12 words per minute morse tests. The Society also produces morse instruction tapes.

RSGB Books

The Society publishes a range of books for the radio amateur and imports many others. RSGB members are entitled to a discount on all books purchased from the Society. This discount can offset the cost of membership.

Technical and EMC Advice

Although the role of the Society's Technical and Publications Advisory Committee is largely to vet material intended for publication, its members and HQ staff are always willing to help with any technical matters.

Breakthrough in domestic entertainment equipment can be a difficult problem to solve as well as having licensing implications. The Society's EMC Committee is able to offer practical assistance in many cases. The Society also publishes a special book to assist you. Additional advice can be obtained from the EMC Committee Chairman via RSGB HQ.

Planning Permission

There is a special booklet and expert help available to members seeking assistance with planning matters.

Send for our Membership Information Pack today and discover how you too can benefit from these services. Write to:

RADIO SOCIETY OF GREAT BRITAIN,
Lambda House, Cranborne Road,
Potters Bar, Herts EN6 3JE

CUT ALONG DOTTED LINE

RADIO AMATEURS' EXAMINATION MANUAL (14th edn)

We hope you found this book really useful in your studies. Please let us have your comments and suggestions for the next edition so we can make it even better!

Name.. Callsign...................................

Address...

....

....

....

FOLD 1

AFFIX
STAMP
HERE

RSGB Book Editor
Radio Society of Great Britain
Lambda House
Cranborne Road
POTTERS BAR
Herts EN6 3JE

FOLD 2

SEAL WITH ADHESIVE TAPE HERE